CURIOSITIES OF HAMPTON ROADS

CURIOSITIES OF HAMPTON ROADS

GHOSTLY COLONISTS, HIDDEN CRYPTS, THE BLACK SWAN OF WESTOVER AND MORE

TAMY KAY THOMPSON

THE History PRESS

Published by The History Press
Charleston, SC 29403
www.historypress.net

First published 2015

Manufactured in the United States

ISBN 978.1.62619.988.0

Library of Congress Control Number: 2015942730

For my husband, Dave—my chauffeur, computer technician and biggest supporter.

CONTENTS

INTRODUCTION

Welcome to the Hampton Roads area of Virginia, which refers to the southeasternmost tip of the state, east of the Fall Line where the land is flat and the brackish waterways are affected by the tide. To the west of the Fall Line, an invisible vertical line running alongside I-95, the terrain steadily becomes more mountainous, the first waterfalls can be seen and the salty water turns fresh. Hampton Roads includes but is not limited to the Southside (Virginia Beach and Norfolk), the peninsula (Newport News and Hampton), Gloucester and the Historic Triangle (Jamestown, Williamsburg and Yorktown).

Virginia, and Hampton Roads in particular, is truly America's birthplace: eight different U.S. presidents were born in Virginia, the first permanent English settlement was established in Jamestown and three major American wars were fought in Hampton Roads. Virginia is named after Queen Elizabeth I (1533–1603), who was known as the "Virgin Queen of England" (and Ireland) because she never wed. The flag of Virginia, adopted in 1861, is the obverse side of the state seal and depicts the personified virtues, called Virtus, with a bare left breast (the only nudity among the fifty seals). She is crushing Tyranny (Great Britain) underfoot, surrounded by the motto *Sic semper tyrannis* ("Thus always to tyrants"). These facts concerning Virginia's early years sum up the state perfectly.

The history of Virginia is steeped in folklore, legends and exaggerated tales, sometimes making it difficult to find an accurate version of the truth. But in all of the different versions, one undisputed fact emerges—the history of Virginia is a fascinating one. On our upcoming tour, you will travel through eight cities and visit over thirty historical sites, including several haunted houses, a hotel, a mental institution, a howling cave, a cursed tree and, of course, a few ancient cemeteries. You will encounter tales of piracy, the first American conspiracy theory, a falsely accused witch who was acquitted three hundred years too late and the final resting place for the first African American born in America. You will even search for buried treasure, learn about the first American fairy tale and walk in Pocahontas's footsteps. Along the way, you will be treated to some beautiful scenery, too, with the bluest skies and the most picturesque beaches. As Captain John Smith wrote of Virginia in 1612, "Heaven and Earth never agreed better to frame a place for man's habitation."

I

THE SOUTHSIDE

We will start our tour of the Southside (the Virginian cities of Hampton Roads south of the James River) in Virginia Beach, the longest beach resort city in the world; Virginia's tidal shoreline measures 1,200 miles in length. Though not ideal for surfing, the riptides on the Southside can be deadly just the same. Strong currents can pull a swimmer or body-boarder offshore in a matter of minutes. A study from 1998 to 2008 revealed that a total of thirty-three people have drowned at the two most popular beaches on the Southside, Oceanfront in Virginia Beach and Ocean View in Norfolk.

And riptides aren't the only danger at the beach—there is also the occasional shark attack. Since the mid-1950s when the International Shark Attack File began, a dozen shark attacks have been recorded in Virginia. One of the most memorable Southside attacks occurred in 2001 when a ten-year-old boy was attacked while wading in four feet of water with his father. Unfortunately, the boy died shortly after from the injuries he sustained.

Since it was established in 1634, Virginia Beach has become a great vacation spot with delicious food, trendy shopping and thriving nightclubs. It also offers fishing, boating and swimming in the Atlantic Ocean with access to the Chesapeake Bay, which is largely considered a dead zone today due to the lack of oxygen in the water. The Southside has also been home to TV evangelist Pat Robertson (*700 Club*) and musician Clarence Clemons (Bruce Springsteen and the E Street Band). Of course, the Southside has its share of historical sites, such as the beautiful Cavalier Hotel, located at 4201 Atlantic Avenue in Virginia Beach.

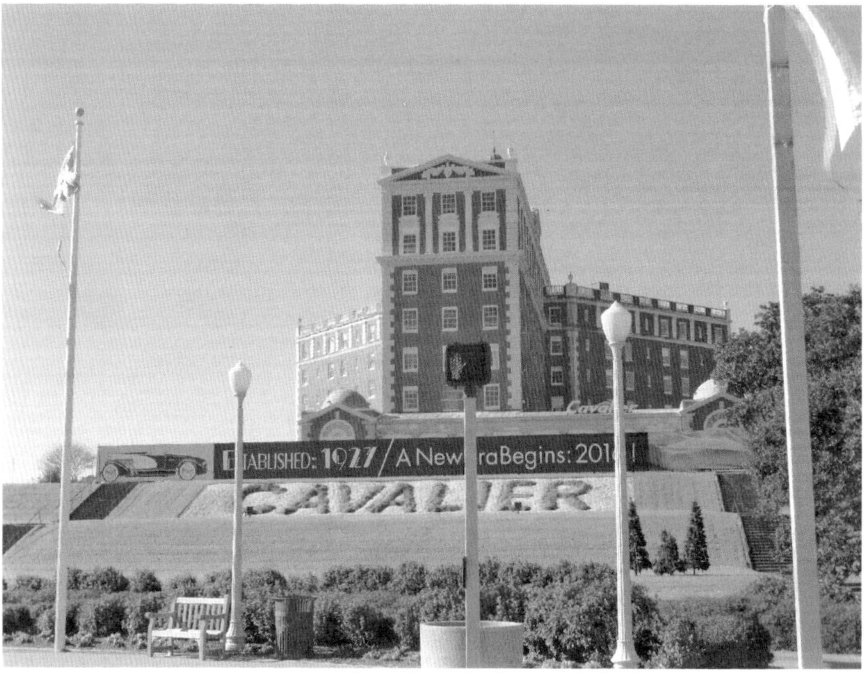

The Cavalier on the Hill. *Photo by author.*

Cavalier Hotel

The original hotel, situated atop a hill, was made from cement-encased steel, and then half a million bricks were added to the structure. The hotel was completely fireproof before the term "fireproof" was even coined. It took about a year to build, opening in 1927 as "the building made from the most bricks" in Virginia. To increase tourism, a train ran straight from Ohio and Chicago to the Cavalier. Aptly, the train was called the Cavalier, and its motto was "Take the Cavalier to the Cavalier" right up until the train's disuse in the 1940s.

There was no electric refrigeration system when the hotel originally opened; only large blocks of ice to keep perishable items cold. Likewise, there was no running water as we know it; instead, there was a large tub of ice on the roof of the hotel. The ice would melt and run down gravity-powered pipes into the hotel, so the water running out of the faucets was only as warm as the sun's rays could make it. The bathtubs were only equipped to run cold salt water directly in from the ocean, making for a frigid bath during the winter months.

The hotel featured luxurious amenities for its time, such as the swimming pool on the roof, which was filled with fresh seawater until the 1970s. The hotel lobby contained a doctor's office, a barbershop and a photographer's studio for the guests' convenience. Plus, there was a stock brokerage office with a ticker tape directly from the New York Stock Exchange. A radio station (WSEA), the third in America to broadcast coast to coast, was located in the hotel as well.

Many famous people have stayed at the Cavalier since it opened, such as Adolph Coors, the original beer tycoon himself, who vacationed at the Cavalier Hotel in the summer of 1929. Coors managed to make it through the early years of Prohibition by manufacturing ceramics, inventing "near beer" and providing malt to the Hershey's Company for its malted chocolate balls. He may have made it through Prohibition, but he couldn't make it through the Great Depression; instead, he plunged headfirst from his sixth-floor balcony at the Cavalier Hotel and died on June 5, 1929.

Throughout the 1930s, the Cavalier was the place to stay during vacations and holidays. Celebrities such as F. Scott Fitzgerald, Ginger Rogers, Frank Sinatra and Judy Garland all stayed there, as well as Richard Nixon, John

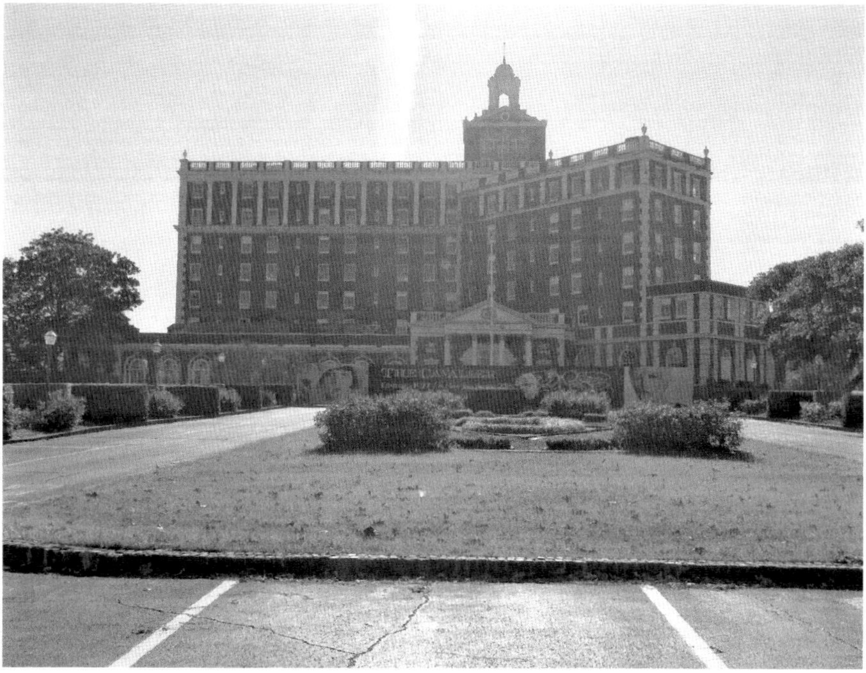

The side entrance of the Cavalier. *Photo by author.*

F. Kennedy, five other United States presidents and a Nobel Peace Prize recipient. Even part of an NBC TV-movie was filmed there in 1992 about the life of Jackie Kennedy, the late president John F. Kennedy's wife.

During World War II, the navy used the hotel as a radar training facility, and the swimming pool on the roof was drained and used as a classroom. In 1944, a naval officer was attending the school when he met his future wife, who was working there as a nurse. The couple soon married. They planned to spend their forty-seventh wedding anniversary at the Cavalier Hotel. Unfortunately, the husband died in 1991, three months short of said anniversary. As a tribute, his wife decided to stay at the Cavalier without him. She gave the hotel manager the flag from atop her husband's casket, and it was proudly flown from the roof of the hotel for a brief time.

A second hotel, the Cavalier Oceanfront, was opened in 1973, and the original hotel was renamed the Cavalier on the Hill. The Cavalier Oceanfront, which is located just across Atlantic Avenue from the Cavalier on the Hill, is less grand and more modern than its original counterpart but is a great place to stay during summer vacations due to its views and proximity to the beach. The Cavalier on the Hill was scheduled to close and be auctioned off but was reopened in 1976 by popular demand. In 2014, the Cavalier on the Hill was closed again for renovations until 2016, and some contents from both hotels were sold.

The Cavalier on the Hill has quite a reputation for being haunted, despite its renovations. Former employees have heard eerie music and the sound of flushing toilets coming from empty sixth-floor rooms, and the front desk clerks have seen calls from vacant rooms light up the switchboard. The elevator sometimes runs by itself, stopping on various floors to open its doors and allow invisible guests to get on or off. Invisible fingers have been known to play the piano in the grand ballroom.

The Cavalier isn't the only eerie place in Virginia Beach, though. We will continue traveling on Atlantic Avenue to our next destination, just off Shore Drive. But along the way, we will pass by Edgar Cayce's Association for Research and Enlightenment on the corner of Atlantic Avenue and Sixty-seventh Street. Since the establishment of the Cayce Hospital in 1929, numerous paranormal research groups have sprung up around Hampton Roads, such as the Hampton Roads Paranormal Society in Gloucester, the Paranormal Activity Research Alliance in Newport News and the Virginia Paranormal Investigations in Hampton; but Edgar Cayce was the original supernatural leader in Hampton Roads.

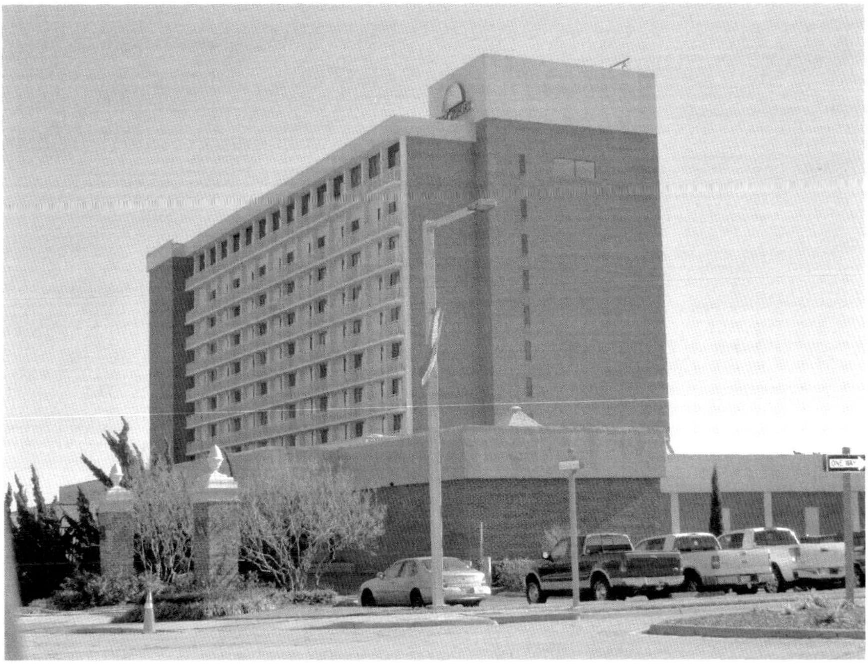

The Cavalier Oceanfront. *Photo by author.*

Edgar Cayce (1877–1945) was a mystic and psychic who gave readings concerning illnesses and holistic treatments while in a trance, and Woodrow Wilson and Thomas Edison were among his list of clients. Cayce was born in Kentucky and, in 1925, moved to Virginia Beach, where he spent the remainder of his life. The Cayce Hospital was established in 1929 but closed shortly thereafter because of the Great Depression. Edgar Cayce died from a stroke in 1945 and was buried in Kentucky.

Ten years later, in 1955, the Association for Research and Enlightenment (ARE) bought the hospital to use as its headquarters. Renovations took place during the summer of 2014 under the direction of the foundation president, Charles Thomas Cayce, Edgar Cayce's grandson. Currently, the ground level of the original hospital is a health center and spa, which offers massages, chiropractic alignments, hypnotherapy, acupressure and much more. Many alternative-style conferences are offered at the hotel each year, mostly about hypnotherapy, mediumship and spiritual awakening. The visitor center offers lectures, a meditation course and even extrasensory perception (ESP) testing. The building also contains a metaphysical library that houses fourteen thousand of Cayce's psychic readings, which were first

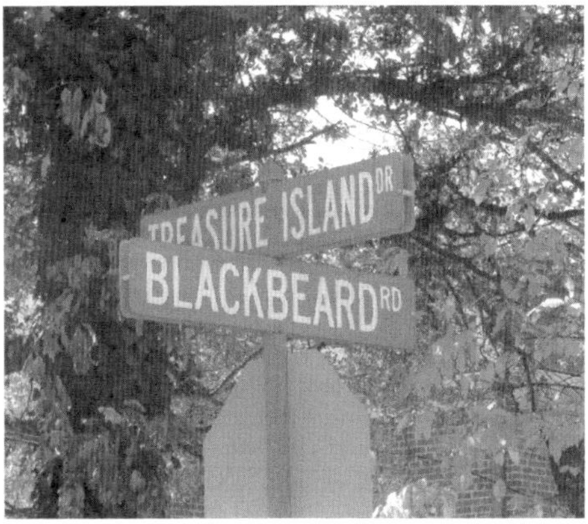

The Treasure Island Drive and Blackbeard Road street signs. *Photo by author.*

recorded in 1923. The facility is located at 215 Sixty-seventh Street and is an interesting place to visit, although spa treatments are by appointment only.

Despite Cayce's supernatural abilities, he never seems to have searched for Blackbeard's lost treasure like we are about to do. Prepare yourself for pirates, because our next stop is Treasure Island, a small island on Lake Joyce next to a residential neighborhood off Blackbeard Road (coordinates: 36.9124, -76.1210). According to legend, this island is where the pirate Blackbeard was forced to abandon some valuable buried treasure.

TREASURE ISLAND

Blackbeard's real name was formerly believed to be Edward Teach, but recent evidence suggests his true name was Captain James Beard. Edward Teach was an alias he used, and Blackbeard was simply a nickname that eventually took on a life of its own. Blackbeard was the most dreaded pirate of the high seas during the early 1700s, despite the fact that he was only a pirate for two years. He was from North Carolina—not England, as he claimed—as was most of his crew. They turned to piracy for profit and fun.

Blackbeard stood six feet tall and had bushy, black hair and a full beard, which he braided with colorful ribbons. He was married fourteen times, mostly to child brides, and was absolutely ruthless in battle. At the height of his career, he had a crew of nearly four hundred men and four ships at his disposal, including the infamous *Queen Anne's Revenge*. This ship ran aground off the coast of North Carolina in 1718 but was discovered in 1996 and confirmed in 2011. Cannons were recovered from it in 2013. When Blackbeard was attacking other ships, he flew his flag depicting a horned skeleton holding a goblet in one hand and preparing to stab a bleeding heart with the other on a solid black background. He also fastened slow-burning lit fuses made from hemp into the hair along the edge of his hat during battles. He completed his ensemble by arming himself with a cutlass that could reportedly split a man in half, several hidden daggers and six pistols crisscrossed on his chest in specially made holsters.

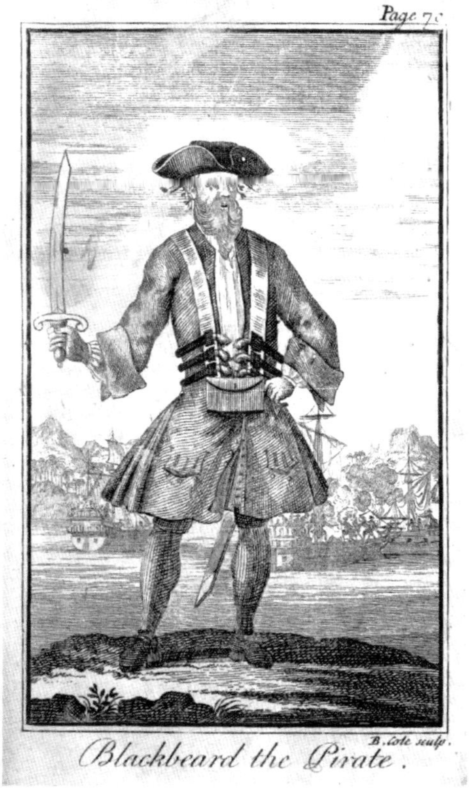

Blackbeard the pirate. *Photo courtesy of Wikimedia Commons.*

Like most pirates, Blackbeard's ties to Virginia were purely through debauchery, mostly attacking and plundering ships as they entered the Chesapeake Bay. He and his crew would steal cargoes of slaves and tobacco—two commodities as valuable as gold—and sell them elsewhere. By 1755, more than forty-two million pounds of tobacco had been exported from Virginia in large barrels known as hogsheads, and most ships exiting the Chesapeake Bay were laden with the product. This made Blackbeard's job as easy as taking candy from a baby.

His home base, and where he retired for a short time and was pardoned, was the eastern shore of North Carolina. He only came north to Virginia to loot other ships and to relax at a local "pleasure house," a tavern with prostitutes. While Blackbeard availed himself of the amenities the pleasure house had to offer, some of his crew would act as lookouts on a nearby dune, known today as Pirate's Hill. The crew would keep an eye out for incoming ships to attack or for patrolling government ships from which to flee.

On one such visit to the pleasure house, Blackbeard's lookouts spotted an incoming ship brimming with riches, and they sent word to Blackbeard. He and his men attacked it. A few of the attacked ship's crew members boarded a small boat and attempted to escape with a chest filled with

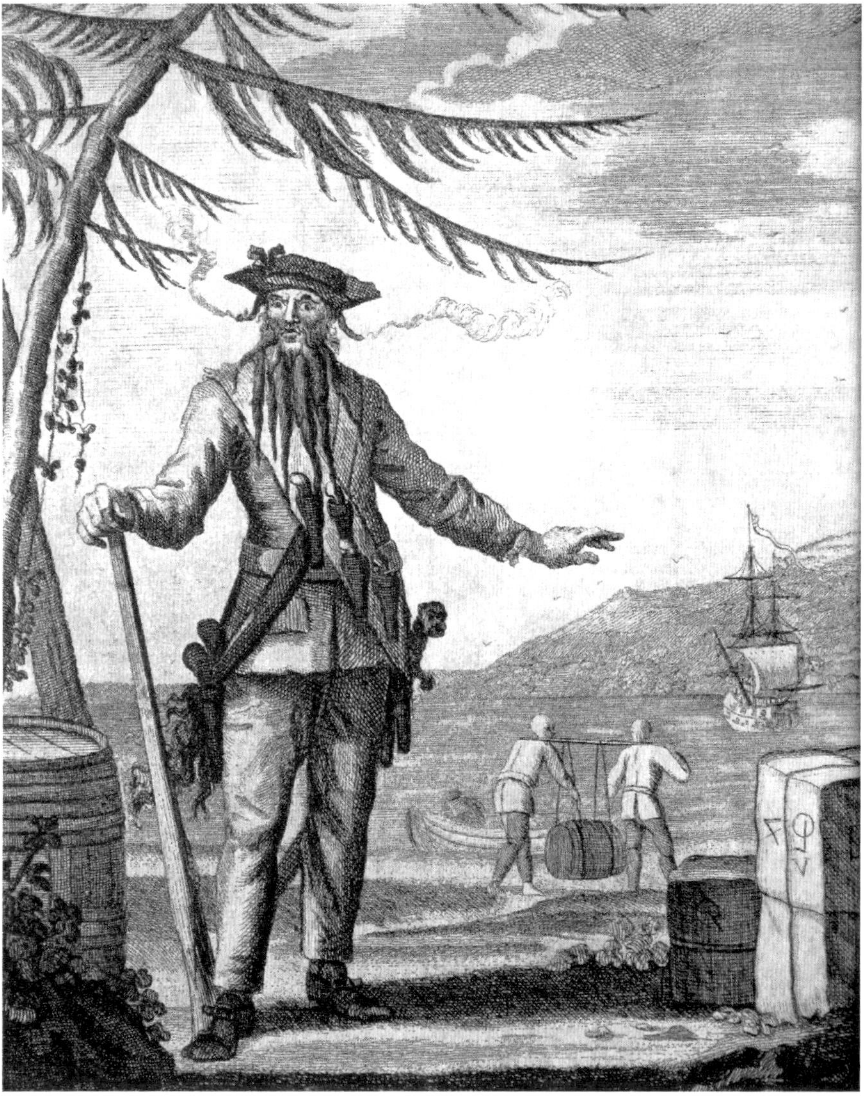

Blackbeard and his booty. *Photo courtesy of Wikimedia Commons.*

valuable treasures. However, Blackbeard spied them, and he and two of his men jumped aboard the small boat and killed them.

Suddenly, two navy ships approached out of the darkness, always intent on capturing or killing any pirates they encountered—especially those working with Blackbeard, since Governor Alexander Spotswood had sent soldiers to capture him. Blackbeard was forced to row ashore to the nearest piece of land. On the small island of present-day Lake Joyce, Blackbeard hastily scratched out a hole in the sand, buried the treasure chest and planned to return at a future time to retrieve it.

Instead, Blackbeard was beheaded by Captain Maynard of Hampton on November 22, 1718. Blackbeard was approximately twenty-eight years old when he died. More than half of his crew also perished during the battle, but many of the survivors who surrendered were returned to North Carolina unharmed. In January 1719, two members of Blackbeard's crew were hanged in Hampton and buried facedown. The remains of one were found in the 1980s and are currently part of an exhibit at the Hampton History Museum. Other crew members stood trial in Williamsburg; one was acquitted, one was pardoned and the other six were sentenced to hang. In 1992, archaeologists discovered a large scaffold designed for mass hangings, indicating that the sentence was carried out on the remaining crew members. In all likelihood, their bodies were displayed along the banks of the Hampton River. Blackbeard's remains suffered a similar fate; his piked head was sailed from the prow of Maynard's ship before being posted in the Chesapeake Bay

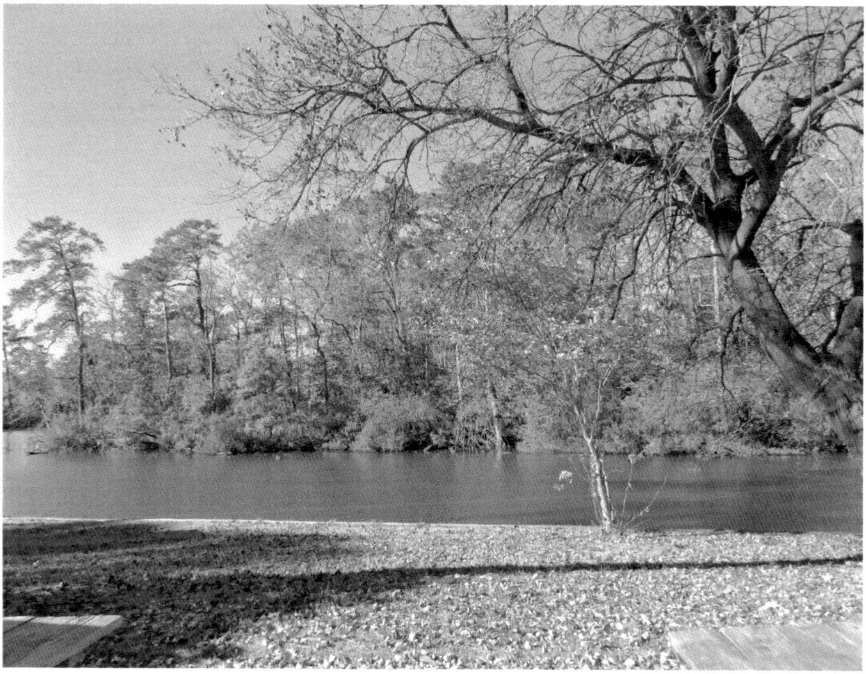

Treasure Island on Lake Joyce. *Photo by author.*

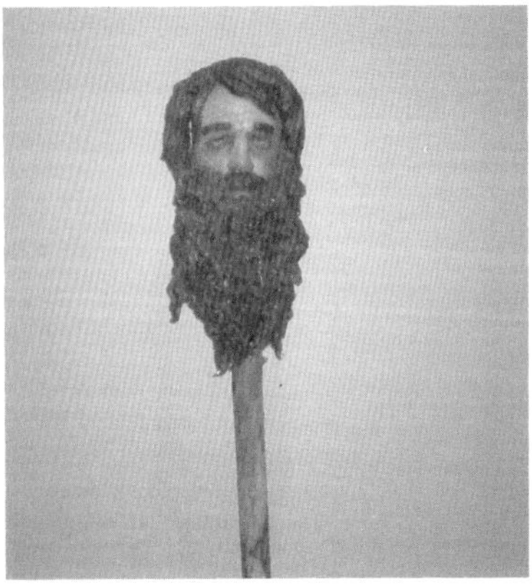

Blackbeard's piked head. *Photo courtesy of the Hampton History Museum.*

at the mouth of the Hampton River for years. This particular spot on the bay is still known as Blackbeard's Point, and a Blackbeard festival is held in downtown Hampton every summer.

Blackbeard's skull was eventually removed from its pike at Blackbeard's Point only to be plated in silver and used as a drinking vessel similar to a punch bowl at the Raleigh Tavern in Williamsburg. From there, no one is quite sure what happened to the skull, just like no one has ever found Blackbeard's buried treasure on Lake Joyce. Many treasure hunters—mostly locals and residents of the adjacent neighborhood—have tried to find it over the years, but no one has ever succeeded.

You won't find Blackbeard's Point or Treasure Island on a map, but there are plenty of other documented tributes to Blackbeard on the Southside. There is Pleasure House Road, Treasure Island Drive and, of course, Blackbeard Road. In fact, you can take Treasure Island Drive to Shore Drive before turning onto Independence Boulevard to our next destination, the Lynnhaven House.

Lynnhaven House

The Lynnhaven House is located at 4401 Wishart Road, just off Independence Boulevard. This charming cottage near the Lynnhaven River has been known by a variety of names over the years, including the Wishart House, and is situated on 5 of its original 250 acres. The Boush family cemetery is on the grounds. Some restoration took place on the house in the 1970s, but over 80 percent of the house is original.

The property was first established in 1648 by Saville Gaskin and then was sold in 1662 to William Hodge, who in turn sold it to Francis and Abigail Thelaball in 1721. They built the current house around 1725; the original carpenter's chalk marks can still be found on the ceiling of the first-floor hallway. Frederick Boush bought the property and house from the Thelaballs in 1789, and it remained in the Boush family until 1859, when George and

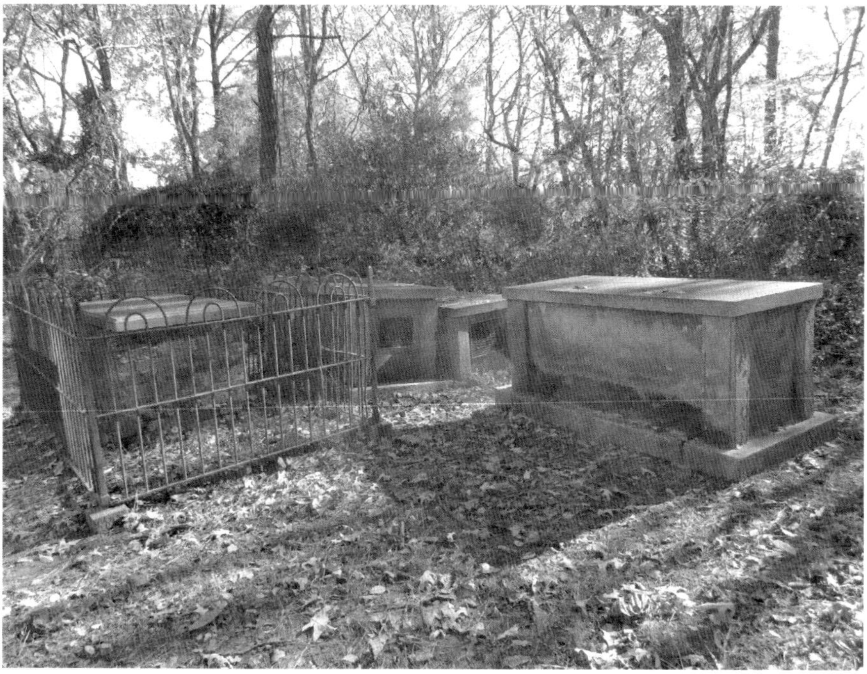

The Boush cemetery at the Lynnhaven House. *Photo by author.*

Joseph Smith bought it. From 1861 until 1922, the deed changed hands a total of seven more times before ending up in the Braightwell family. In 1923, the Olivers bought it from the Braightwells, and they became the last private owners of the Lynnhaven House. As of 1971, when the Olivers donated the house to the Association for the Preservation of Virginia Antiquities, the house still didn't have electricity or running water. Currently, the City of Virginia Beach operates the house, and it is listed in the National Register of Historic Places.

In 2007, the original timbers in the cellar and attic were drilled in an attempt to determine the exact age of the house. To date, the results of said test revealed that the wood was cut in the fall of 1724, suggesting the house was built in 1725. But this doesn't explain why a previous archaeological excavation turned up a piece of broken glass with the date 1713 etched onto it. Perhaps the owner of the property in 1713, William Hodge, planned to build a house and transferred a windowpane from his former house to his new property. The house was never built, though, and the window sat unused until it was inadvertently broken.

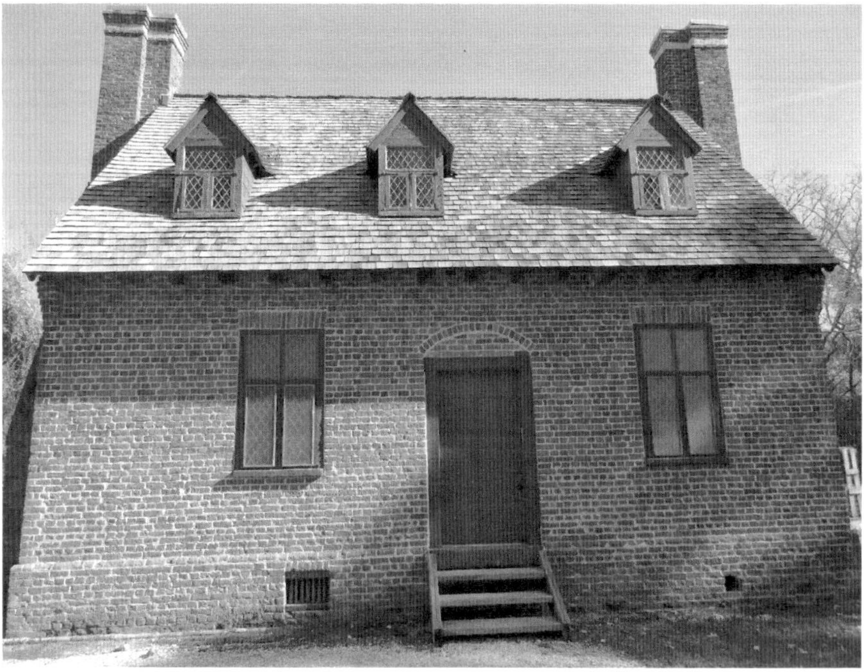

The Lynnhaven House. *Photo by author.*

Tourists and tour guides of the Lynnhaven House have seen a trio of ghosts in the house: the spirits of a man, a woman and a little girl in a dazzling white dress. Civil War reenactors, while camping out on the property overnight, saw a ghost standing in the nearby family cemetery. Also, a workman who was restoring sections of the house kept misplacing his tools only to discover they had been hidden around the house by invisible hands. The Olivers, the last people to inhabit the house, had their fair share of supernatural events occur in the house, too. They reported hearing the sound of footsteps walking around overhead, although no one was on the second floor at the time. The Olivers also witnessed doors and cabinets opening and closing by themselves, and they saw all three ghosts at different times while living there.

The Lynnhaven House is a well-known landmark, making it a popular tourist attraction and very crowded at various times of the year. Our next stop is usually crowded, too, but for different reasons. The Sentara Bayside Hospital at 800 Independence Boulevard is not only a well-equipped medical facility, but it is also where a tribute to Grace Sherwood, the infamous Witch of Pungo, can be found.

Grace Sherwood Memorial

Grace White was born in the mid-1600s in a small community of Virginia Beach then known as Pungo. The community sat on present-day Indian River Road and was named after the Chesapeake Indian chief Machlopungo. Grace married James Sherwood in 1680, and her father gave them 50 acres of land as a wedding present (he left his remaining 145 acres to them in his will). The couple had three sons: John, James and Richard. Grace was knowledgeable about the medicinal properties of herbs, and she preferred wearing men's trousers while picking herbs and roots, although her memorial statue at the hospital depicts her in a dress. She was a trusted local healer and an experienced midwife who had delivered many of Pungo's children.

Grace lived in a time when witchcraft was a serious offense and to be publicly accused of witchcraft could end in death. In 1697, one of Grace's neighbors, Richard Capps, privately accused her of being a witch when his cow stopped producing milk. Although the accusation wasn't made in public, it stung Grace, and she retaliated by suing her neighbor for defamation of character, one of the first such cases in America. Grace's case was soon dismissed, but her troubles were far from over.

A year later, Grace filed two more lawsuits against her neighbors. Both neighbors had accused Grace of witchcraft prior to her lawsuits. One neighbor, Jane Gisbourne, stated that Grace had killed her crop of cotton by bewitching it; the other neighbor, Elizabeth Barnes, claimed Grace and Satan had sneaked into her bedroom at night to bedevil her and her husband. Supposedly, Grace escaped their bedroom by turning into a cat and squeezing through the keyhole! Grace lost both defamation lawsuits against the neighbors, and she was ordered to pay the court costs. Thankfully, she wasn't found guilty of witchcraft, or she would have lost more than a little bit of money—she could have lost her life.

Grace's husband, James, died in 1701 and left all of his earthly possessions, including the house and property, to Grace. She barely had time to grieve his passing before suing yet another of her neighbors, Luke and Uxor Hill, in 1705, this time for assault and battery. Surprisingly, Grace won this particular case, along with twenty shillings, perhaps because the bruises were still visible on her skin.

Grace's victory was short-lived, though. Only a short time later, those very same neighbors publicly accused Grace of witchcraft. Grace was summoned to appear in the county court where her body was carefully searched by matrons, or older married women, for "witch's marks." A witch's mark could be anything from a birthmark, a wart or an oddly shaped freckle to a mole, a melanoma or a discolored splotch of pigmentation. One of the matrons, and the jury foreman, was none other than Elizabeth Barnes, whom Grace had previously sued. So it was no wonder when two different witch's marks were found on Grace's body: "We of the jury have searched Grace Sherwood and have found two things like tits with several other spots."

When news of her witch's marks reached the public ear, accusations around town began to fly. One rumor claimed that Grace could fly to England and back at night in an eggshell. This ridiculous misconception stemmed from an innocent response Grace had given years earlier after being asked where the rosemary in her garden had come from. Grace explained how the herb had traveled from England in an eggshell, a common way of transporting seedlings in a ship, and the townspeople chose to believe that Grace could fly in a magic eggshell at night.

After the case was kicked around various courts, it eventually returned to the county courthouse, where the judge decreed a trial by witch ducking on June 7, the first such trial in the state of Virginia. Witch ducking is when the accused witch is thrown into deep water to prove her innocence or guilt; if she sinks, she's innocent but dead, and if she swims, she's guilty and dead just the same. Witch ducking was the ultimate lose-lose situation for someone accused of witchcraft.

Each of Grace's thumbs was tied to the opposite big toe, and then she was rowed in a small boat into water above her head in an area now called Witchduck Point. There, she was tossed overboard to either sink or swim. Grace chose to swim and promptly bobbed to the surface, thus proving she was a witch. A preacher aboard the boat attempted to push her back under water by pressing her with his Bible.

She survived the ordeal, and her life was spared, but she was found guilty of witchcraft and imprisoned for several years, losing her property in the process. Grace was eventually released—but not pardoned—and given her house and land back after she paid the back taxes. She died of natural causes in 1740 at the age of eighty. She left five shillings each to her sons James and Richard; everything else was given to her son John. Upon her death, her fellow townspeople were so afraid of her spirit returning in the form of a cat, an animal she'd been particularly fond of in life, that the men killed every pet and stray feline they could find. This unnecessary "catricide" directly led to the great rat infestation of

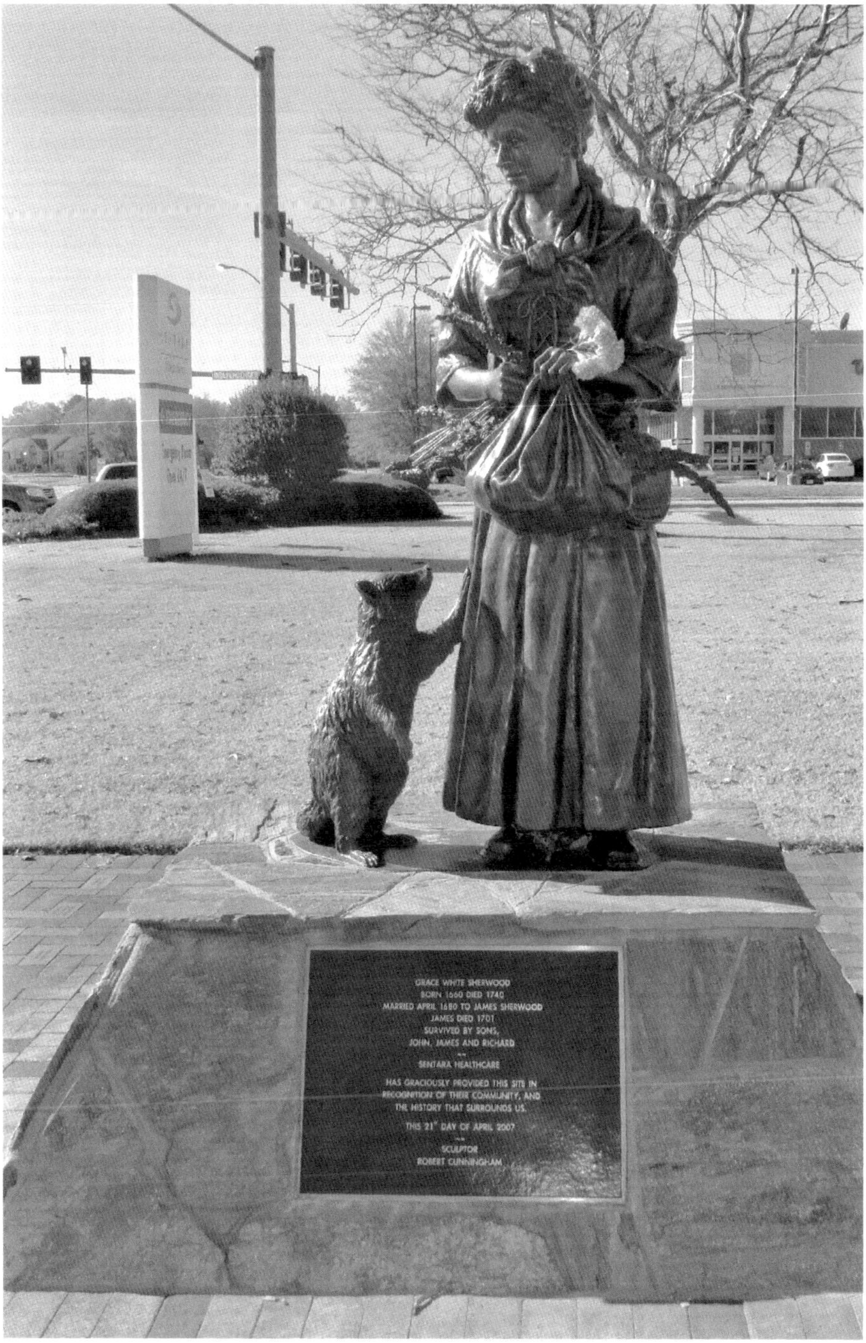

A statue of Grace Sherwood. *Photo by author.*

Virginia Beach in 1743 and fueled the belief that cats could be used as witch's mediums.

Grace Sherwood is now known to have been kind to animals, knowledgeable about herbs and an experienced midwife. She was also the only citizen of Virginia to be convicted of witchcraft due to a trial by witch ducking. Exactly three hundred years to the hour of Grace's witch ducking, Governor Timothy M. Kaine finally pardoned her in 2006—a little too late, some would say. A memorial statue and plaque, which calls Grace "one of our first healers," are located at the Sentara Bayside Hospital. Other tributes to Grace can be found in Virginia Beach as well: Witchduck Road, Sherwood Lane and a spot on the water known as Witchduck Point.

The memorial to Grace Sherwood is our last stop on the Southside, and we will need to get back onto Shore Drive in order to go to the peninsula. But Shore Drive has a story all its own. Shore Drive is only a little over three miles long, but it is perhaps the deadliest stretch of highway in America. The road was completed in 1928, but the land beneath the road has a much older history.

SHORE DRIVE

Shore Drive's bloody history began in 1638 when three hundred dead shipwreck victims washed ashore on this particular stretch of land. It was a popular spot to execute pirates well into the 1700s. There is also a mass Native American burial site that was discovered there in the 1970s. The early Algonquian Indians of Hampton Roads didn't bury their dead in the same fashion as the English, making the mass burial site very curious indeed. It's believed the site contains the corpses of the Indians who were mass murdered in an event you will learn more about in Chapter 7.

Perhaps this treacherous stretch of road is cursed, which would explain why so many pedestrians, bicyclists and motorists have died on it over the years. There have been so many deaths, in fact, that a plaque keeping track of the death toll was hung on Shore Drive in 1977. The sign hung at the foot of the Chesapeake Bay Bridge Tunnel, and eighty-three deaths were recorded by 2004, around the time the sign was removed. Since its removal, even more deaths have occurred on Shore Drive, including a local reporter who was writing about the dangers of the road at the time of his unfortunate death.

We will continue on Shore Drive to the peninsula, but be sure to keep to the posted speed limit and drive carefully until reaching our next destination: Newport News. It is the first of many stops on the peninsula and one that is least traveled by tourists, making it an ideal city to explore. There, you will find an early church, a couple curious buildings and more than one obscure cemetery.

2

NEWPORT NEWS

To get from the Southside to the peninsula, you will have to traverse bridges and tunnels in order to cross the Chesapeake Bay. When you arrive on the peninsula, you'll be greeted by the cities of Newport News on your left and Hampton on your right. Newport News is recognizable by the looming skeletal structures of the shipyard and Hampton by the languishing military base called Fort Monroe. The tip of the peninsula is divided, with the northeastern side being the city of Hampton and the southwestern side the city of Newport News; we will enter the peninsula via the Hampton Roads Bridge Tunnel to the western city. Both cities were discovered by the English captains John Smith and Christopher Newport when they explored the James River and surrounding areas in 1607. But the city of Hampton was settled long before the city of Newport News.

Christopher Newport, captain of the *Susan Constant*, led and was solely in charge of the overseas voyage of the English expedition to America. Upon landing at Jamestown on May 13, 1607, the settlers began working on a fort. Captain Smith began exploring the surrounding countryside and waterways, while Newport ventured up the James River and adjoining rivers for a few weeks. Captain Newport, per his employers' instructions, was looking for a water route to the Pacific Ocean. Finding none, he returned to England to report to his employers, the Virginia Company of London, and to gather more supplies to bring back to Virginia. Captain Newport quickly set off for America again but arrived at Jamestown later than he had planned, not landing until January 1608. He entered Virginia through the Chesapeake

Bay to the entrance of the James River, and this point soon became known as "Newport's News" because he not only brought provisions from England, but he also brought news from the mother country.

But Captain Newport was worthy of having a city named after him long before the Virginia Company of London employed him. He was an experienced seaman who had fought under Sir Francis Drake against the Spanish Armada. After that epic battle, Newport began plundering silver-bearing Spanish ships in the name of England, and he eventually lost an arm. He was a privateer (a glorified pirate, though legitimate) and looted Spanish ships in the Caribbean for thirteen years before joining the Virginia Company of London and becoming part of early American history. He's honored in Virginia through the city of Newport News and Christopher Newport University, which has a statue of its namesake. Currently, Newport News is perhaps best known for having one of the best shipbuilding yards in the world. It is also the birthplace of several famous people, including the singers Pearl Bailey and Ella Fitzgerald and the author William Styron.

However, the Newport News side of the peninsula struggled much longer than the Hampton side, which has flourished since its founding in the early 1600s. What finally brought life to Newport News was the Chesapeake and Ohio (C&O) Railroad. It laid its tracks through the city in 1859, with a single train making the run once a day, and trains use the same track to this very day. But the city itself lacked life; the train tracks were about the only thing in Newport News at that time, and it remained so until Collis P. Huntington, president of the Old Dominion Land Company and known as the "robber baron from California," arrived in town. Huntington knew there was money to be made in building an empire, and he was just the person to single-handedly finance the endeavor.

Newport News didn't even have an official name when Huntington arrived, boasting a mere one thousand residents who mostly worked for the railroad or on private fishing boats. The majority of residents lived together in one area on what is now Eighteenth Street and River Road, then referred to as "Hell's Half-Acre," which adequately described the living conditions.

The unnamed town had a grand total of nine small buildings—a couple saloons, a chapel without a minister, a couple general stores and a few other businesses—when Huntington arrived. Upon Huntington's arrival, the Old Dominion offices were added. The town lacked a courthouse and jail, and the C&O construction chief also served as the justice of the peace. There were two doctors in town to treat the many accidents associated with fishermen and railroad workers, and they had the two most bustling businesses in town.

There weren't enough white children in town to warrant a school; however, there was an African American schoolroom, simply known around town as "the colored school."

Huntington's arrival changed the entire town in more ways than one. He envisioned Newport News as an affluent place to permanently reside and a vacation resort for summer tourists. He achieved his dream by building the framework of present-day Newport News, including shops, businesses and hotels. All of Warwick County, which encompassed the city of Newport News at one time, was permanently altered by Huntington's vision.

Around the same time that Newport News was officially named a city in 1881, ships began to dock at its convenient shore for repair work. Huntington was never one to pass up an opportunity, and he soon built the Newport News Shipyard, which was opened for business in 1886. The shipyard got its first navy contract in 1893, thus establishing itself as "the best ship-builders in America." It also launched the most haunted ship in history, the ill-fated USS *Hornet*. Incidentally, the shipyard is still in operation today as one of the largest in the United States but is now called Northrop Grunman Newport News Shipbuilding. But it always has been and always will be known simply as "the shipyard" to the local Hampton Roads residents.

Hotel Warwick

Huntington also built the Hotel Warwick, which is our first stop in Newport News. The Warwick is located at 2410 West Avenue across from the World War I memorial Victory Arch, on the corner of Twenty-fifth Street and West Avenue in downtown Newport News, near what used to be a nice pier leading to the James River. The hotel opened in 1883, and John Swinerton ran it after its grand opening. The hotel was originally made from wood; it was a luxurious four-story building, intended to accommodate the wealthiest midwestern and Yankee vacationers, claiming it was "the best hotel south of Washington." Located inside the hotel for the convenience of its upper-class patrons was the office of the first newspaper, *The Wedge*, and the first bank in Newport News.

Fifty years later, the hotel was deemed a fire hazard and torn down. Ironically, despite being rebuilt out of brick, most of the hotel was destroyed by fire in 1960. Only a portion of it remained intact and was still in use until 1988. In 1991, the City of Newport News converted the hotel into

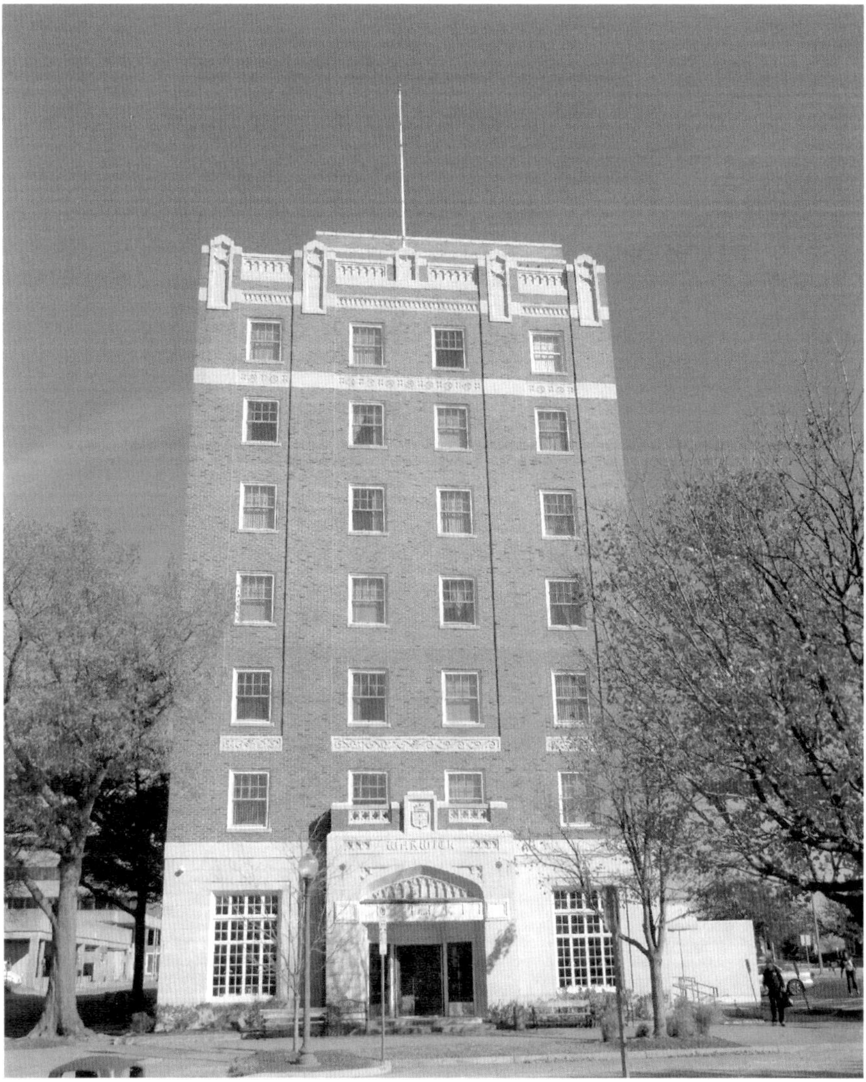

The Warwick. *Photo by author.*

single-room occupancy (SRO) apartments for the homeless. Funding for the conversion was received from banks, the state government and private sources. Huntington's dream of an elegant vacation spot became eighty-eight inexpensive apartments for transitional unfortunates. The Warwick SRO apartment building opened its doors in 1995 and was filled to capacity within three months.

Wilson Family Cemetery

We will leave the Warwick, as most of its tenants do, and travel west on Route 60, or Warwick Boulevard, to the Wilson Family Cemetery. Along the way, you will catch glimpses of the shipyard; go through quaint Hilton Village, where Jackie Tomlin investigated a haunted house in 2010; and pass by Christopher Newport University. The Wilson Family Cemetery is located off Deep Creek Road near the Hidenwood area of Newport News, near Todd Stadium. The Wilson family originally owned the property on which the cemetery stands from at least the late 1700s until the early 1800s, when the Parker family acquired it.

This privately owned cemetery is among the oldest in Newport News, but few people, including neighboring residents, have ever heard of it. It's situated atop a hill near the spot where the original house stood; few trees are left to provide any shade. There are no aboveground tombstones or headstones, but old graves are sometimes visible, depending on the conditions of the overgrown weeds and underbrush.

The Wilson Family Cemetery was described in the Virginian Historical Inventory as: "Location: 4.3 miles south of Denbigh, Virginia, on Route #60; thence, 1.3 miles west on Deep Creek Road; thence, 0.1 of a mile on a private road; thence, 0.3 of a mile west on private road; thence, 1 mile north on private road." It takes luck to find it according to these directions, but it's well worth it once you finally track it down. The cemetery is worth visiting just to see its ghosts, if nothing else. Various neighbors, friends and family members knowledgeable about the cemetery have seen misty apparitions in the old graveyard…and not just when it's foggy outside, either.

Denbigh Baptist Church

Our next stop in Newport News is close by, just off Route 60 and Oyster Point Road, at the Denbigh Baptist Church, located at 13010 Mitchell Point Road. It was built in 1774 and was the third church built for Warwick Parish, making it the oldest building in what was formerly Warwick County (currently the city of Newport News). The church is one of the few surviving colonial structures in Newport News and is one of the oldest Baptist churches in America.

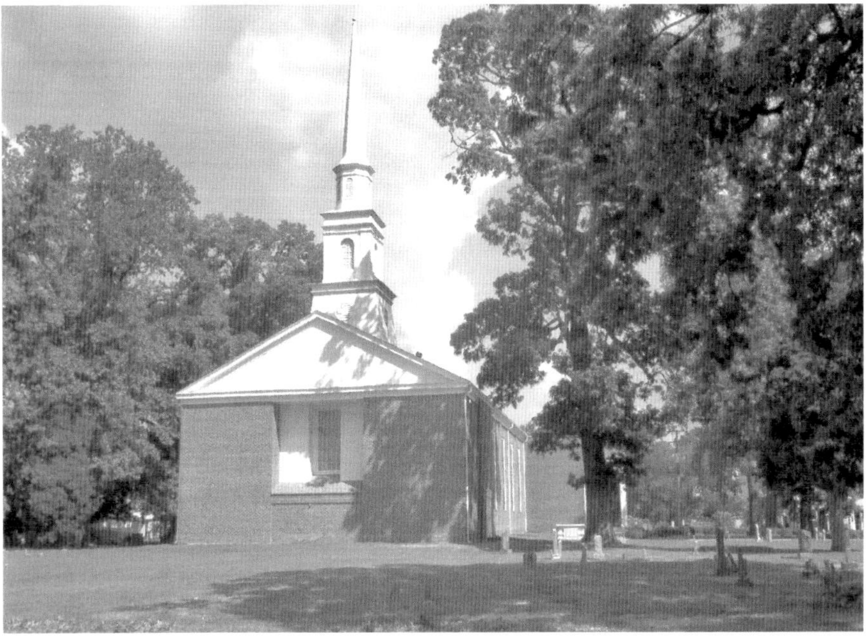

Denbigh Baptist Church. *Photo by author.*

The first church of Warwick Parish was built in 1627 and housed a congregation of Episcopalians. This church was destroyed, and its foundation bricks were used to help build the second church, located slightly north of the current building. The second church was built around 1662 and still housed a dwindling number of faithful Episcopalians. Soon even this small number stopped attending, and the church was temporarily abandoned.

The third and last church of Warwick Parish was built in 1774 and was no longer owned by Episcopalians due to the lack of practicing members. Over the next twenty-five years, several different religious sects owned the church until the Baptists took it over in the early 1800s. In 1810, it was renamed Warwick Church. The church, still in Baptist hands, was severely damaged during the Civil War; the interior was utterly destroyed, but oddly, the exterior suffered very little harm. After the war, African Americans began holding services in the abandoned church but were eventually forced to build their own, the First Baptist Church of Denbigh. In 1882, the church was renamed yet again, this time to its current name, the Denbigh Baptist Church. "Denbigh" refers to Denbigh Plantation, a nearby plantation in Warwick County first patented by Samuel Matthews during the first half of the 1600s.

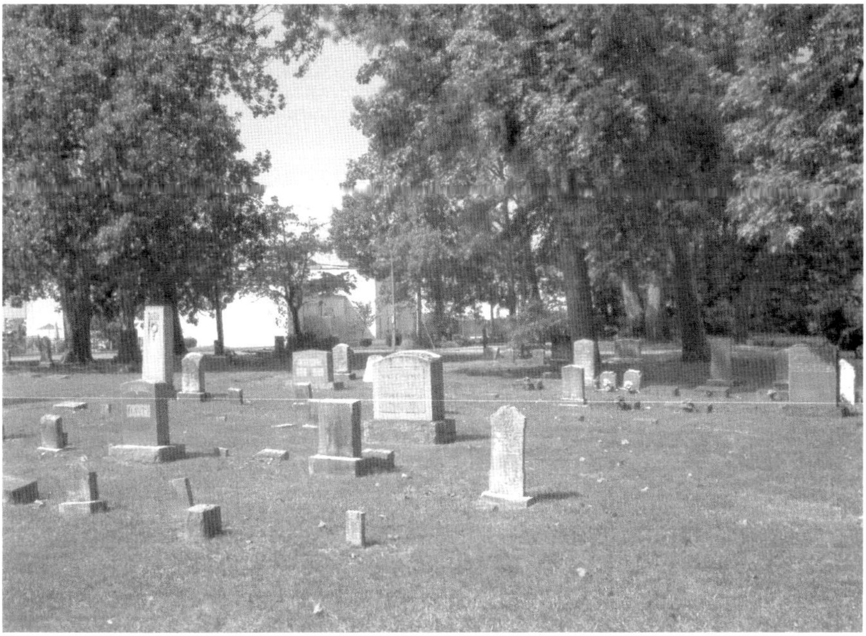

The cemetery at the Denbigh Baptist Church. *Photo by author.*

Denbigh Baptist Church was first remodeled in 1898 when the building was raised from its original single story to its present Gothic design and the square-topped windows were elongated. In 1939, the church was enlarged, and the current parsonage and tabernacle were built. Little else has been done to the church since then, and it looks much like it did at the turn of the twentieth century. The surrounding cemetery is every bit as undisturbed and beautiful as the church, with a single upright monument: the grave of navy surgeon and University of Virginia professor Dr. Bennett Green (1835–1913).

Potter's Field

We will continue heading west on Route 60 to another old cemetery, a bona fide paupers' cemetery called Potter's Field, one of the first in America. This cemetery for the less fortunate is located next to the New Apostolic Church, with its own cemetery containing another fifteen graves, just down the street from the Old Warwick Courthouse.

This particular section of Newport News is known as Denbigh, part of the area originally called Warwick River Shire but renamed Warwick County in 1643. Denbigh Plantation, mentioned in the preceding Denbigh Baptist Church section, was located close by, and this area was the county seat throughout most of the 1800s. Warwick County was named an official city in 1952. In 1958, it consolidated with, and took the name of, Newport News. Since then, Denbigh has become known by local residents as an area in the northern portion of Newport News.

Potter's Field was first used in the early 1800s, purchased by Warwick County in 1893 and then transferred to Newport News in 1896. Originally,

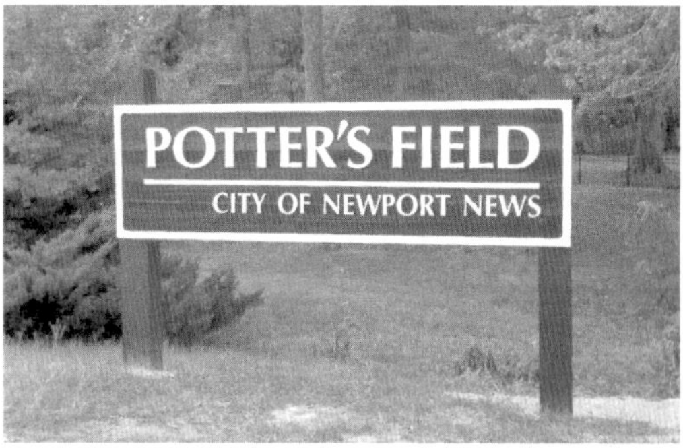

Right: The Potter's Field sign. *Photo by author.*

Below: The Potter's Field memorial stone. *Photo by author.*

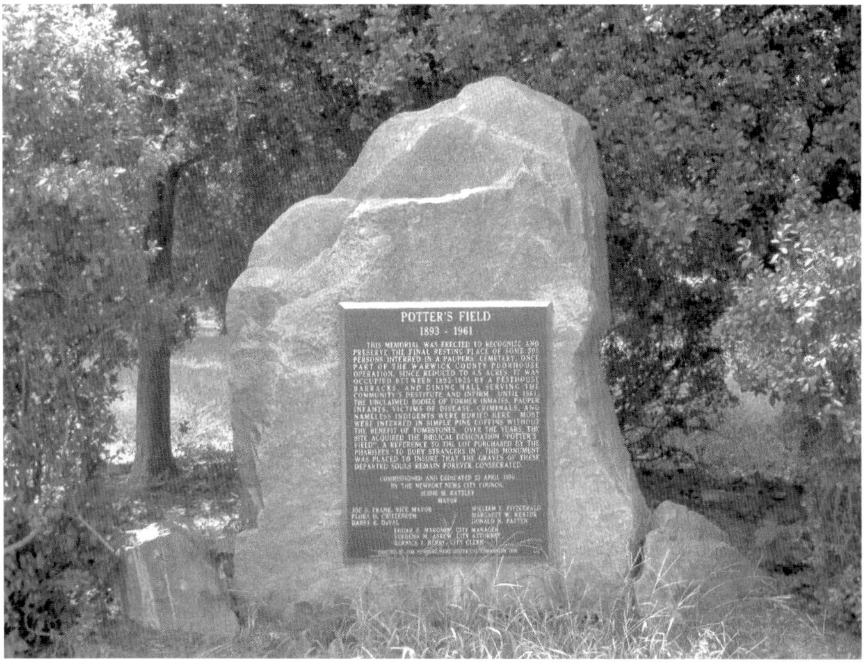

the site was part of the Warwick County Poorhouse Operation; from 1893 until 1935, there stood a dining hall that fed the less fortunate. But it was always a cemetery, used frequently until the 1930s for criminals, unclaimed inmates, "pauper infants," victims of disease and nameless indigents. Most of the arrangements were handled by W.J. Smith Funeral Home, which is still a family-owned business in Hampton Roads today. Approximately five hundred people are interred at Potter's Field, and though they are all unmarked graves, the names of the identified deceased are recorded. The last identified person to be buried there was Mary Carey in 1961.

Perhaps the most tragic cause for an interment at Potter's Field was that of "an unidentified black man" who was lynched in Newport News on June 11, 1902. He wasn't the first African American to be lynched in Hampton Roads nor would he be the last, although Virginia had the fewest number of lynchings out of all the southern states—only one hundred recorded lynchings took place between 1880 and 1930. In 1881, William Allen was accused of killing a white man, and he was pulled out of his jail cell and hanged by a mob next to the Warwick Courthouse. On October 4, 1995, the day after O.J. Simpson was acquitted, Antwan Sedgwick was found hanged by his belt from a set of monkey bars at a playground in Hampton. Local police ruled it a suicide, HBO did a program about it in 1998 and the FBI investigated it in 2000. To date, it is still unclear whether it was a suicide or a lynching.

BOXWOOD INN

Let us now leave this sad place and even sadder subject. We will continue traveling west through the Denbigh section of Newport News until we come to the charming Lee Hall Village. Here, we will visit a beautiful bed-and-breakfast named the Boxwood Inn, located at 10 Elmhurst Street. Simon Curtis opened the Boxwood Inn in 1896, a mere fifteen years after Newport News was officially named a city. Its accommodations were only rivaled by the Hotel Warwick, though the Boxwood Inn competed on a much smaller scale.

Simon Curtis worked at the Lee Hall train depot, and he happened to overhear two businessmen talking about how Warwick County needed a hall of records and there was money to fund the project. Curtis approached the men and managed to convince them that he was the man for the job.

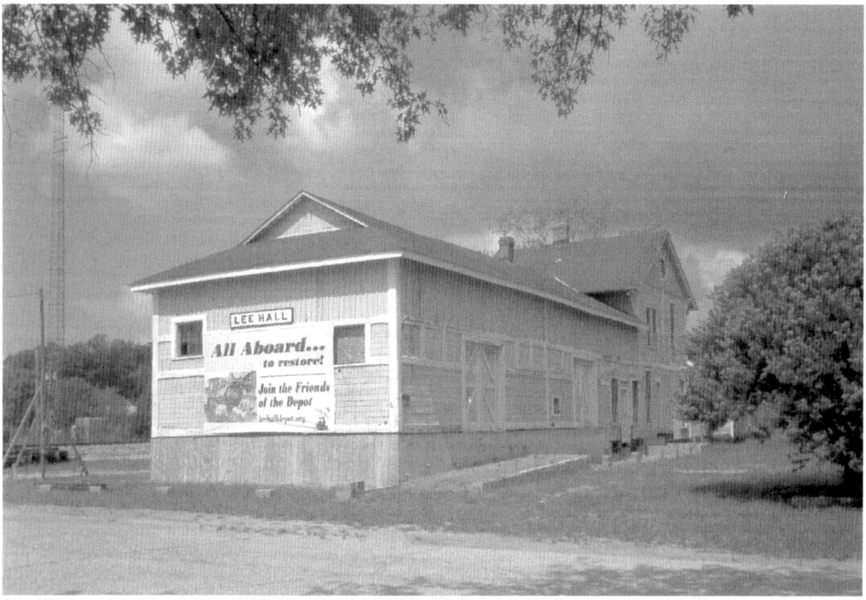

The train depot. *Photo by author.*

Surprisingly, the men agreed to finance Curtis's proposal, and the mansion was built in 1897 with a hall of records located on the first floor.

Shortly thereafter, Curtis added a post office and a general store to his home, and he soon became known as the "Boss Man" of Warwick County. He and his wife, Nannie, lived in the mansion for years and first began renting out rooms to soldiers and their families during World War I. Eventually, Mr. and Mrs. Curtis died, and as a result, their businesses fell on hard times, and the mansion was put up for sale with the contents auctioned off to the highest bidder. A fast-food restaurant put in a bid on the unfurnished building, planning to tear it down and rebuild, but Barbara Lucas bought it in 1995 with visions of restoring it to its former glory and making it a bed-and-breakfast.

Lucas immediately began experiencing strange incidents. On one of her first days in the inn, she broke a fingernail while unpacking a box and wished aloud for a nail file. She looked across the room a moment later and was amazed to see a brand-new emery board sitting on the windowsill. Later, well after its grand reopening, Lucas decided to host a Civil War reenactment ball at the inn. The event went off without a hitch, with one notable exception: an uninvited guest, dressed in a Confederate uniform, showed up. No one knew the gentleman, but it didn't stop him from striking up conversations

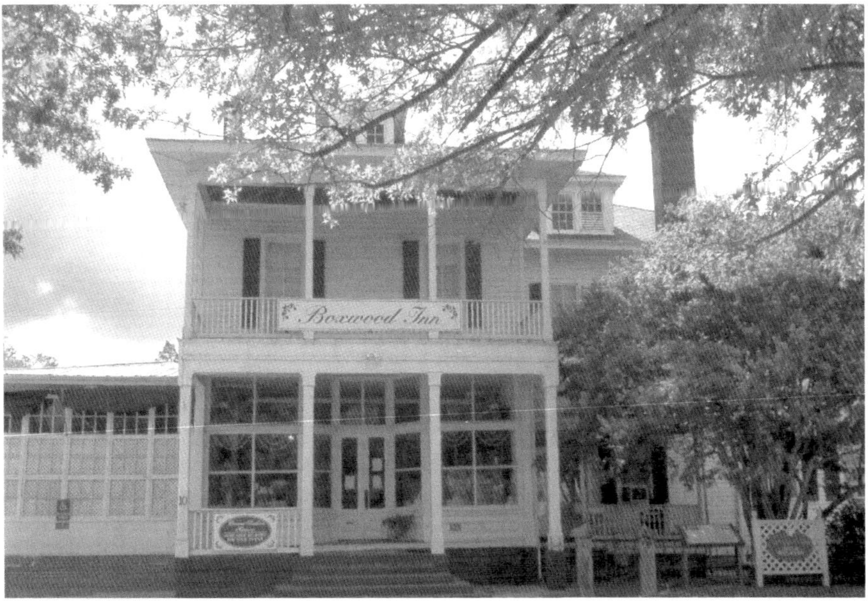

The Boxwood Inn. *Photo by author.*

with several of the guests. During the course of the evening, a photograph was taken of the dapper stranger, but only a streak of light showed up when the film was later developed.

The former owner Nannie Curtis is also supposed to haunt the inn. She enjoys knocking on guests' doors early in the morning as a supernatural wake-up call. Nannie has even been known to open and close random guests' doors, especially lovebirds or newlyweds. Her favorite time to make mischief is when the inn is at its busiest, like Valentine's Day, says current owner Kathryn Hulick, who bought the inn in 2008. A year later, in the summer of 2009, the inn was listed on the National Register of Historic Places and the Virginia Landmark Register.

We will say farewell to Nannie and continue traveling along a northwest course until we come to the first point of the Historic Triangle that we will visit. Jamestown is a tiny island just off the Virginia peninsula and is truly America's birthplace. The first American business was also established (and subsequently failed) at Jamestown. And the first American fairy tale—the love story between Pocahontas and John Rolfe—took place at Jamestown.

3

JAMESTOWN

Jamestown, named after King James I of England, was the first permanent English settlement established in America. The Virginia Company of London financially backed the expedition to America, and Captain Christopher Newport led it. Captain Newport also steered the *Susan Constant*, John Ratcliffe was the captain of the *Discovery* and Bartholomew Gosnold steered the *Godspeed*. Over one hundred men boarded the three ships less than a week before Christmas 1606, but winds delayed them for six weeks. They reached the islands of the West Indies off the coast of South America in late March 1607, when the first Englishman died. George Percy recorded that Edward Brookes's "fat melted within him by the great heat and drought of the country."

On April 26, 1607, the three ships landed at Cape Henry, and that night, the Indians chased the men back to their ships. Gabriel Archer and another man were wounded. A sealed box was opened that day, the contents of which will be discussed momentarily. Shortly thereafter, Captain John Smith was released; he had been shackled below deck on Captain Ratcliffe's ship because he'd been accused of mutiny during the voyage (he and Edward Wingfield had argued about the route Newport was taking). A few days later, the settlers erected a cross at Cape Henry.

On April 30, the settlers began traveling up the James River to look for a settlement site, and they feasted with the Indians at Kecoughtan (present-day Hampton) for the first time. Maize bread and pipes of tobacco were shared, and then the settlers were treated to singing and dancing before

Captain Newport gave the Indians beads from England, which were much appreciated because the Indians liked adornments. Before the English had arrived, the Indian men decorated their ears and animal-hide clothing with animal bones. They also inserted feathers into the left side of their long hair (the right side was shaved) and painted their bodies red or black.

The settlers continued exploring the James River until the night of May 13, when they stopped at the three-mile-long, one-mile-wide island of Jamestown because no Indians lived there. The next morning, on May 14, 1607, the council members were sworn in, and Edward Wingfield was elected president. Almost immediately, the settlers began building a fort.

Over one hundred men made the journey to America, but only thirty-eight survived to see Newport's first supply ship arrive in January 1608. Some of the men came to America for adventure and some to start a new life, but they were all under strict instructions from their backers—they were "to dig, mine, and search for all manner of mines of gold, silver, and copper"; to find Walter Raleigh's lost colony of Roanoke from 1587; and to convert the native Indians to Christianity, among other things. The sealed box that was opened at Cape Henry detailed the settlers' numerous responsibilities in two separate documents.

The first was a list of instructions, assumed to have been written by Reverend Richard Hakluyt, mostly concerning the native people the settlers would encounter. First, the settlers were told to choose their settlement location based on how healthy the natives looked, making sure the site wasn't surrounded by trees in order to avoid being ambushed. The settlers were not to offend the natives because they needed to trade with the natives. They should also build up a store of food before planting any crops so the natives wouldn't realize they meant to stay. Next, the settlers were never permitted to let the natives borrow their weapons and were to allow only the best marksmen to shoot so the natives would fear them. The natives shouldn't know about sickness, death or executions inside the fort. They were told how to build houses in a straight line, how to dig wells for water and other things the gentlemen knew nothing about. The settlers were ordered to not write home about anything negative. Lastly, they were told the way to prosper was to work together.

The second document in the box was a list of the seven names of the governing council members: John Martin, George Kendall, John Ratcliffe, John Smith, Bartholomew Gosnold, Christopher Newport and Edward Wingfield. Wingfield was nominated president, and he recorded in his diary that he was one of three men of the triumvirate. John Martin had previously commanded a ship under Francis Drake and had accompanied Gosnold on a trip to New England in 1602. Bartholomew Gosnold had sailed with Walter Raleigh to New England, where he named Cape Cod and Martha's Vineyard. Gosnold's cousin Edward Wingfield fought against the Spanish Armada and was one of the biggest financial backers of the Jamestown expedition (he rented out most of his properties to acquire the remaining funds needed just two days before setting sail for America). John Smith was a soldier who earned the title of captain and a coat of arms when he beheaded three Turkish soldiers. George Kendall was a Scottish soldier and Sir Robert Cecil's servant. Not much is known about John Ratcliffe other than his real surname was Sicklemore.

Like many governing bodies, the seven men rarely saw eye to eye; a few of them actually despised one another. Kendall's son died during the summer of 1607, and then Kendall accused Wingfield of hoarding liquor and food, so Wingfield had him imprisoned. Bartholomew Gosnold died from malaria in August, and his departure from this earth was honored by a volley of shots, as recorded by George Percy. Shortly thereafter, Gabriel Archer, John Smith and John Ratcliffe deposed Wingfield and appointed Ratcliffe as the new president. While Wingfield was a prisoner aboard one of the ships, Ratcliffe searched his belongings, stole some of his papers and accused him of being an atheist because he didn't own a Bible. Unbeknownst to Ratcliffe, Wingfield's Bible had been stolen days before he set sail for America.

During an argument between Ratcliffe and the blacksmith James Read, Read struck Ratcliffe. In retaliation, Ratcliffe imprisoned Read and sentenced him to death. Once strung up on the gallows, Read named Kendall as a leader of some sort of conspiracy. "Captain Kendall shall die for being a fool Christian," Ratcliffe said before pronouncing Kendall guilty, making this incident the first trial in America. But Kendall objected because "Ratcliffe" wasn't even his real name, so Martin announced Kendall's death sentence. Kendall was executed by a firing squad, and the blacksmith Read was released. Wingfield was also sentenced to death, but Captain Newport returned from his exploration in the nick of time. Wingfield, Ratcliffe and Martin traveled back to England with Newport, but Wingfield didn't return with them to Virginia in 1609.

That is how the first governing council in America began in Jamestown, which became the first capital in Virginia, remaining thus until 1699. Currently, Jamestown Settlement, a colonial period park with reenactors and an Indian village, is why Jamestown is such a popular place to visit. Another popular tourist attraction in Jamestown is the ruins of the first American business, a glass factory, with a newly erected glass studio and glass blowers in period costumes located nearby. The blowers use their mouths to blow glass in front of a large furnace near the ruins of the original glass furnace; they sell their art to the tourists. Neither one of these places is on our tour, but they are both worth seeing nonetheless.

Jamestown Church

What Jamestown is most famous for is its history and legends, including the first American fairy tale: the love story between the Indian princess Pocahontas and the successful tobacco farmer John Rolfe. After a historical romance, the couple were wed in an Anglican church at Jamestown in the spring of 1614, when Princess Pocahontas became Mrs. Rebecca Rolfe. Pocahontas actually had two other names as well—Amonute and Matoaka. Pocahontas was born about 1595; her birth name was Amonute, and her secret name was Matoaka. She kept this last name to herself until adulthood because Indians believed the secret name was magical and could be used against them by their enemies.

The Algonquian tribe, of which Pocahontas was a member, already inhabited southeastern Virginia when the English first arrived in 1607. Almost the entire peninsula fell under Chief Powhatan's domain, with more than thirty individual tribes having their own leaders who, in turn, had to answer to Powhatan himself. Even Pocahontas had to abide by Chief Powhatan's wishes, despite being the favorite of his forty or so children. Since birth, Pocahontas had lived with her mother, but when she turned eight, as was custom, she went to live with her father in his neighboring village. It was Powhatan who bestowed the nickname "Pocahontas" on her because it meant "playful one," and she enjoyed nothing more than entertaining her father by singing, dancing and teasing.

As a typical Indian child, Pocahontas kept most of her head shaved except for a long braid near the back, and she wore little or no clothing. Her hair was shaved about once a month with a sharpened oyster shell, which resulted in a somewhat stubbly appearance. Upon reaching adolescence, Indian children were tattooed, usually on their faces, necks and shoulders but sometimes on their arms, legs or backs. The single portrait painted of Pocahontas during her lifetime clearly shows that she had no tattoos on her face, but the rest of her body is concealed by clothing. In all probability, by the time Captain John Smith met her, Pocahontas had at least one tattoo. A naked Pocahontas taught the English boys how to turn cartwheels at the fort, so the tattoo would have been clearly visible to Smith, who recorded the event.

There were many other differences between the Indians and the English besides their hairstyles, clothing and body art; the Indians bathed more regularly than the English. Likewise, the Indians drank water and rarely drank distilled spirits until the English introduced it to them. The English, on the other hand, thought water was "un-pure," unhealthy and should be ingested only by commoners.

Another vast difference was the Indians' self-sufficiency and their willingness to share this knowledge with the English newcomers. The Indians knew how to live off the land and taught the early settlers how and when to plant and harvest corn, how to smoke meat and cure it with brine and how to fish and extract oysters from the rivers and bay. All they wanted in exchange for this knowledge was some of the loud English weapons to use while hunting deer instead of their bone and wooden arrows. It should be noted that the Indians shared their knowledge with the English settlers even before they agreed to supply them with the coveted guns.

Pocahontas was approximately twelve years old and still considered a child when she first met the charmingly boastful Captain John Smith. He was short of stature and estimated her age as older than it actually was simply because she was taller than him. He also incorrectly assumed, after learning she was Chief Powhatan's daughter, that she was a princess in the traditional English sense of the word. However, Pocahontas would never rule the land because that particular power was matrilineal, or passed down from the mother's side, not the father's. And though Pocahontas's father was the great chief himself, her mother had been nothing more than a pretty Indian maiden who had caught Powhatan's eye.

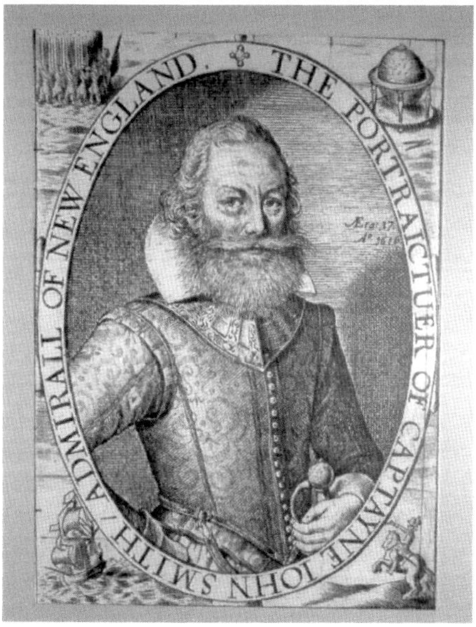

Captain John Smith. *Photo courtesy of the Hampton History Museum.*

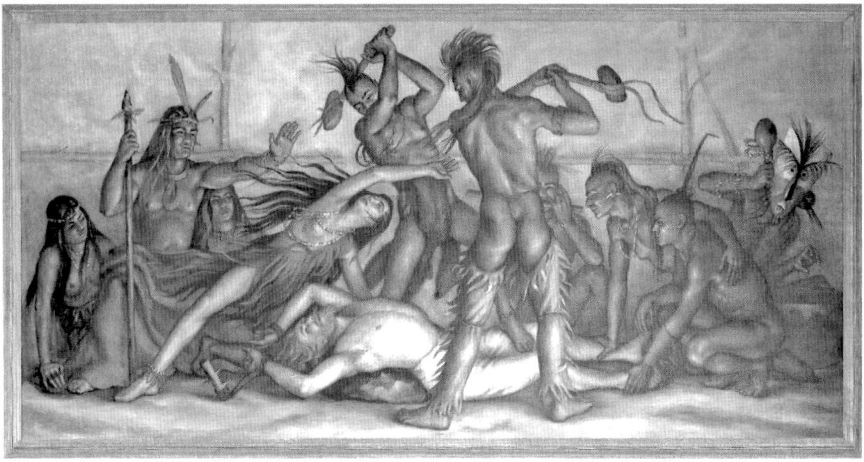

Pocahontas saves the life of John Smith. *Photo courtesy of the Library of Congress.*

Captain Smith was a braggart by nature and, by his own admission, a bit of a ladies' man. One of the first English phrases he taught Pocahontas was the racy remark "Love me not you?" Other examples of his blustery nature include his controversial account of his expedition up the Chickahominy River, where he claimed to have been captured by Chief Powhatan's brother Opechancanough and, under Powhatan's orders, threatened with clubs but was saved by none other than Pocahontas herself. This version of the Chickahominy River expedition was omitted from Smith's original account but was included in the second version he wrote a few years later.

During that same ill-fated expedition, another scout, George Cassen, was also captured by Indians. He was less fortunate than Smith, though. The Indians tied Cassen to stakes and cut off his fingers, toes, arms and legs with sharpened reeds. With only his head and torso left intact, they sliced off his face and scalp, which they hung outside like a flag as a warning to the English. The rest of his remains were tossed on the fire and burned. It was a similar torture technique used on Captain John Ratcliffe two years later in 1609. Ratcliffe was tied naked to a tree, skinned alive, dismembered, gutted and thrown into the fire. Indian women wielded the instruments of torture because they believed Ratcliffe had raped some of their daughters, but Pocahontas was still a child and would have been too young to have participated in the execution.

Pocahontas reached maturity in 1610 and married a fellow Indian from a neighboring tribe in Powhatan's domain. Her husband's name was Kocoum, but the English always referred to him as "Mr. Pocahontas." Pocahontas lived with her husband in his village and rarely visited the fort at Jamestown any longer. Unsubstantiated rumors still persist today that Pocahontas and Kocoum had a child together but that the Indians kept the baby a secret, safely hidden from the English.

Meanwhile, the English settlers were slowly starving to death during the winter of 1609–10. Captain Smith had injured his leg in a gunpowder explosion and returned to England on October 1, 1609, where he faced various charges of misconduct, which you'll learn about in Chapter 5. Captain Smith had been one of the few settlers able to communicate with the Indians, and without his ability to trade (and steal), the settlers soon despaired. They were forced to eat their pets first and then rats, shoe leather and, finally, the fresh corpses of settlers who had recently starved to death. By the end of that long winter, known as the Starving Time, one man even killed his pregnant wife and hid caches of her body parts around his house for later consumption.

The Indians had already shown the English how to feed themselves, yet the English kept demanding that the Indians trade with them. Perhaps the Indians would have been more willing to help if the English had upheld their end of the bargain by providing the weapons, muskets and swords in particular, in exchange for corn. To make matters worse, it was the third straight drought year, and the Indians' crops had suffered, too. Plus, each tribe under Powhatan's domain had to pay him an annual tribute of corn regardless of the drought, and they simply didn't have enough to go around.

By 1613, relations between the Indians and the English had deteriorated to the point that trade became impossible. Chief Powhatan had lost patience with the dishonest Englishmen, but he still wanted their weapons. The Indians began stealing items from the settlers; some even made a game out of stealing weapons while the settlers' backs were turned or stealing an axe out of an unsuspecting settler's hands as he chopped wood. The English lost a number of weapons this way and were truly exasperated by the time Captain Samuel Argall came up with a devious plan involving Pocahontas.

Pocahontas had been traveling around her father's land without Kocoum when the plan was implemented. She was staying at a neighboring tribe, and with Chief Iopassus and his wife's help, she was lured aboard Captain Argall's ship under false pretenses. She was locked in a cabin and taken as a hostage to Jamestown. The English rewarded Chief Iopassus and his wife's betrayal with a single copper pot. Captain Argall incorrectly figured that the chief's favorite daughter was the perfect leverage to get Powhatan to return their stolen weapons and provide enough corn to tide them over. Powhatan agreed, but the few guns he did return were dismantled and rendered completely useless. The English soon realized that Powhatan would never concede, and his daughter was too valuable to harm—if anything untoward happened to Pocahontas, the Indians would very likely start a war. Therefore, Pocahontas remained a prisoner and was encouraged by the English to abandon her heathen lifestyle and become a proper young lady. She attended an Anglican church, where she was eventually baptized and adopted the Christian name Rebecca. Only then did she reveal her secret Indian name, Matoaka.

During her months of imprisonment, the renowned tobacco farmer John Rolfe fell in love with Pocahontas, perhaps because there were so few English women around. It wasn't until 1620 that ninety women were sent to become wives at Jamestown, where they were valued at 150 pounds of tobacco a piece. Regardless, Rolfe asked Governor Thomas Dale's permission for the couple to wed. Dale, in turn, gained Chief Powhatan's reluctant approval before giving his consent. Reportedly, King James I of England was less than thrilled by the match between an Indian princess and a mere commoner, even if he was the most successful tobacco plantation owner in America. Nevertheless, John Rolfe and Pocahontas were married at the Jamestown church on April 15, 1614, and eight years of peace between the settlers and the Indians ensued. Incidentally, the first load of tobacco was shipped to England soon after their wedding.

Rolfe and his new bride had a son whom they named Thomas, presumably after Governor Thomas Dale. Curiously, the baby isn't registered on any documents—baptismal, legal or otherwise—but it's believed he was born in 1615 or early 1616. While he was still an infant, Thomas and his parents set sail for England to visit Rolfe's family. Pocahontas, her husband and their baby stayed in England for ten months, during which time she met Queen Anne and sat to have her portrait painted by Dutch artist Simon Van de Passe. Pocahontas conducted herself like royalty and was introduced as "Lady Rebecca, the Indian Princess."

Pocahontas, Rolfe and Thomas embarked on their return voyage to America in March 1617. Poor weather stalled the ship, named the *George*, but the foul weather made no difference to Pocahontas, who had fallen ill as soon as she boarded. Before the ship could leave the banks of the Thames, Pocahontas died of a respiratory ailment at the age of twenty-two. She was buried at St. George's Church in Gravesend, England, but the exact location of her grave is unknown.

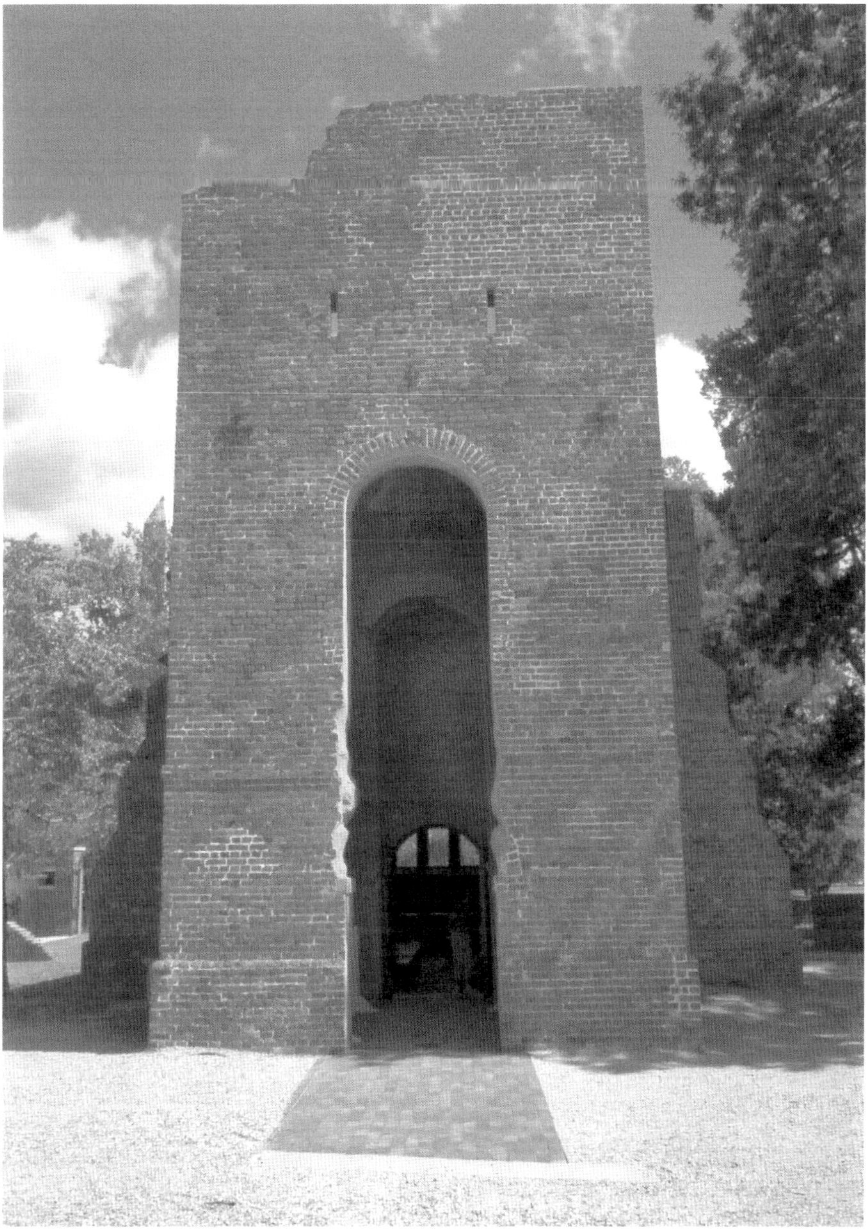

Jamestown Church. *Photo by author.*

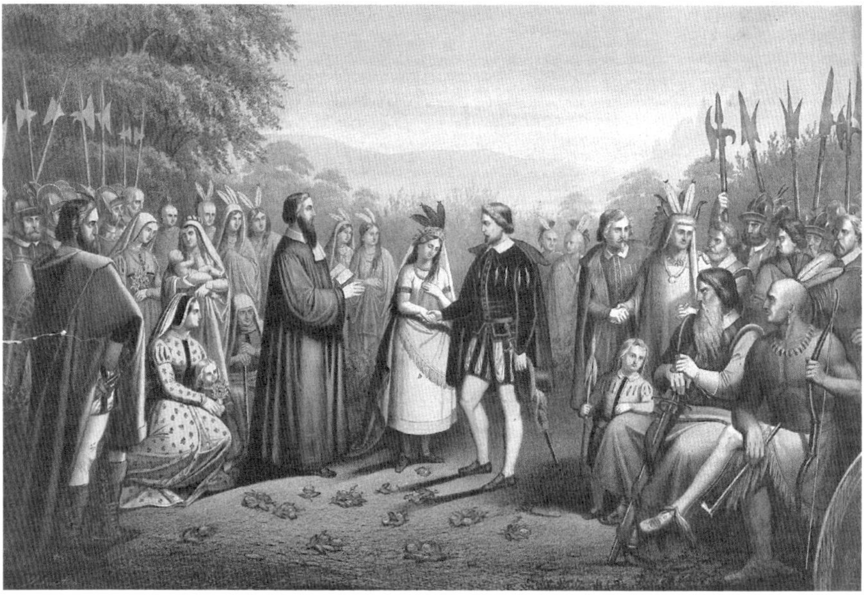

The wedding of John Rolfe and Pocahontas. *Photo courtesy of the Library of Congress.*

John Rolfe resumed his trip home, minus his wife and son, whom he chose to leave in England to be cared for by relatives. Unfortunately, father and son would never see each other alive again. By the time Thomas was an adult and returned to Virginia, both his father and maternal grandfather had died. Chief Powhatan died of natural causes a year after Pocahontas, and John Rolfe didn't survive the Indian Massacre of 1622, led by Powhatan's brother Chief Opechancanough.

The great Indian Massacre of 1622 will be mentioned in upcoming chapters as well, so now is a good time for an explanation of the event. An Indian uprising was a long time in the making, but one event seemed to trigger it. Only months before the actual massacre, the Indian Nemattanew, who sometimes helped the English and sometimes helped the Indians, is believed to have killed a colonist. Then, Nemattanew—who was described by George Percy as wearing feathers and swan wings on his shoulders—was shot by a colonist's field hand. Shortly thereafter, the Indian Massacre occurred on Good Friday and was as finely orchestrated as if it took place during the digital age rather than colonial times. The Indians in southeastern Virginia united on the morning of Good Friday 1622, and at a pre-appointed time, they turned on the closest English people and proceeded to slaughter them. Some Indians had even brought

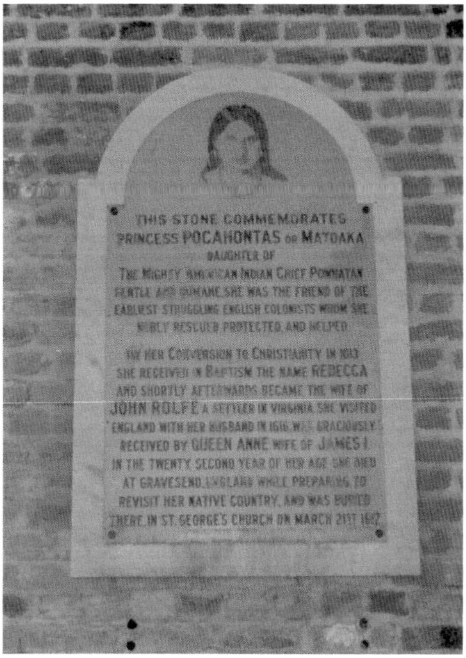

Pocahontas's plaque. *Photo by author.*

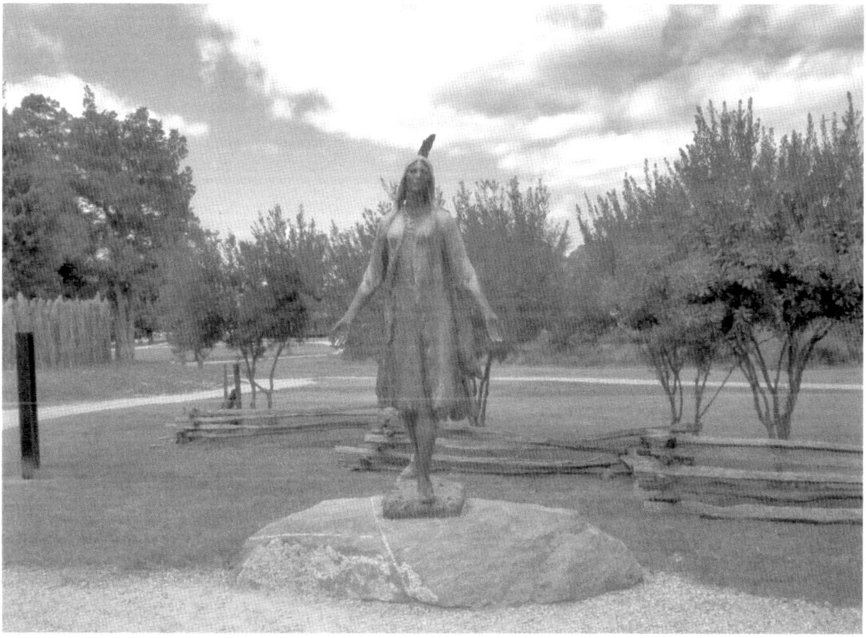

A statue of Pocahontas. *Photo by author.*

much-needed food to the settlers the day before and joined them on Good Friday to share a meal. The Indians then sprung from the tables and killed their English dinner companions, not sparing man, woman or child.

But the Indians weren't satisfied with their bloody handiwork and created even more carnage by hacking at the mutilated corpses with their tomahawks until the dead English were virtually unrecognizable. Most of the survivors buried the scattered remains of their friends and family in mass unmarked graves, lacking the strength and stomach to separate the remains. All in all, a total of 350 English settlers died that day, with John Rolfe being one of the many victims.

This is the sad end to the fairy tale of John Rolfe and Pocahontas. Although the exact location of their human remains is a mystery, statues of both Pocahontas and John Rolfe can be found at Jamestown. A necklace belonging to Pocahontas is on display at the nearby Mattaponi Reservation in West Point, one of the two first Indian reservations in America (the other is the Pamunkey Reservation in King William County). Their son, Thomas, was buried at Kippax Plantation, located in Hopewell, on the south bank of the Appomattox River just off I-295. The land was bestowed on Pocahontas as a wedding present from her father, though she never got a chance to live there. Neither did Thomas, but his wife and daughter are buried there with him. In 2006, the nonprofit Archaeological Conservancy agency purchased the property, acquiring the $205,000 needed with a grant from the Virginia Land Conservation Fund. Currently, tourism is frowned on because the site is being preserved and professionally excavated.

JAMESTOWN CEMETERY

Now, we will visit our next stop at the Jamestown Memorial Church, located about a mile from the Jamestown Settlement park. The current church was built in 1907 atop the original 1617 foundation, and bricks from the old church were used to make a wall around the cemetery. However, it is not the church or the cemetery we're interested in, it is the sycamore tree located on the cemetery grounds. This particular tree is known as the "Curse Tree" and has an interesting legend—concerning the relationship between a meddling mother and her willful teenaged daughter—associated with it.

Sarah was a beautiful seventeen-year-old who lived at Jamestown with her wealthy, plantation-owning parents in the late 1600s. She had just

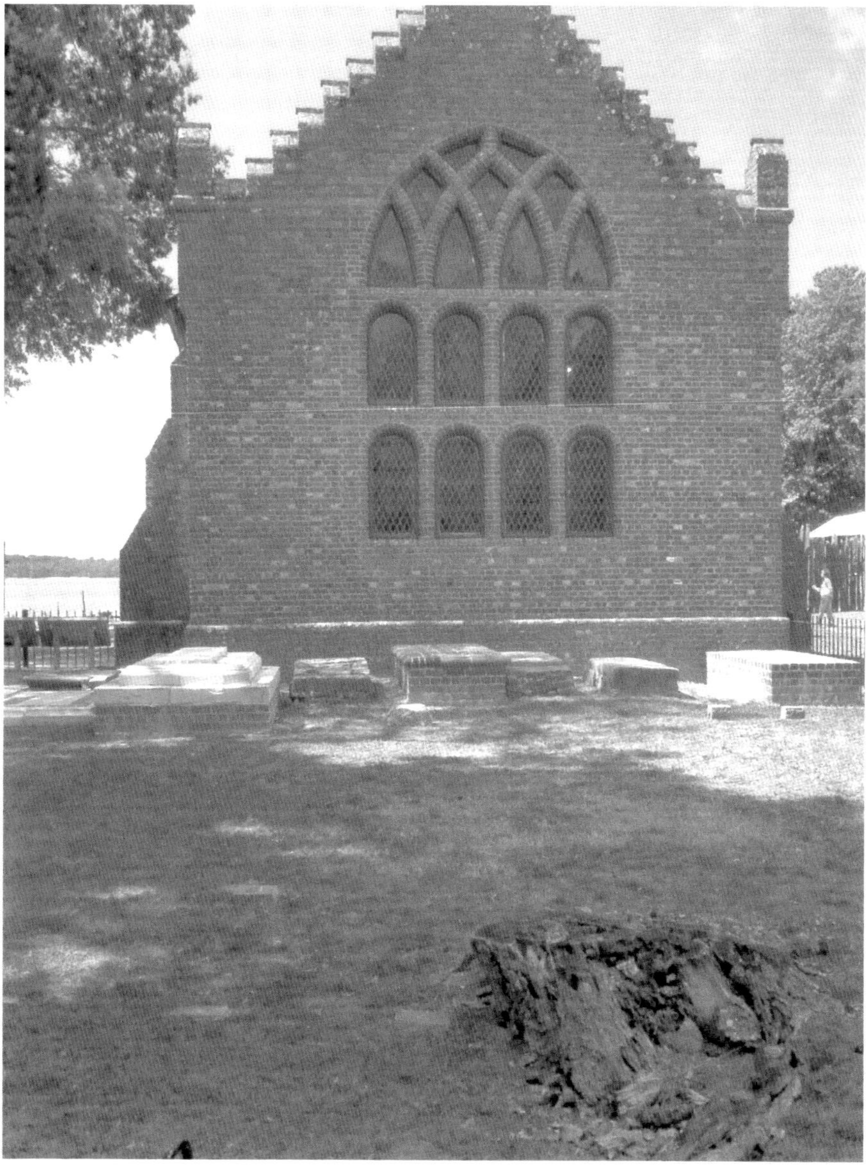

The cemetery at Jamestown Church. *Photo by author.*

signed a marriage contract with an upstanding young man her parents approved of when she met the distinguished Dr. James Blair a mere three weeks later. Dr. Blair was a prosperous preacher of good standing, but Sarah's parents thought he was far too old at the age of thirty-one to be a proper suitor for their daughter. Despite her parents' opposition, Sarah broke her original marriage contract and accepted Dr. Blair's hasty proposal. Her parents refused to attend the wedding, and her mother vowed to never rest until she had kept the couple apart. A short time later, both of Sarah's parents were struck and killed by lightning and were buried at the Jamestown Memorial Church cemetery.

Sarah and Dr. Blair lived happily ever after, until her death in 1713 at the age of forty-three. Dr. Blair lived another thirty years and, upon his death, was buried at the Jamestown Memorial Church, only six inches away from his beloved wife. Soon a sycamore tree sprouted up between the two graves and was left to grow untended. The tree continued to grow to maturity, and eventually, its roots pushed Sarah and Dr. Blair more than seven feet apart. Sarah's body ended up within inches of her mother's grave—a woman who seems to have accomplished in death what she couldn't in life.

We'll leave this historical site of ill-fated love and venture for another—the ruins of Lydia Ambler's family home. At one time, the house was a magnificent Georgian-style mansion, but now only a few partial walls rise forlornly into the air with blue sky where the roof should be. It is situated on Jamestown Island, near the settlement site and Jamestown Memorial Church.

AMBLER MANSION

The island of Jamestown was deserted after the capital of Virginia moved to Williamsburg in 1699. Eventually, the island was split, with half belonging to the Travis family and the other half owned by the Ambler family. Richard Ambler was the collector for the Port of the York River in 1724, and he was married to the heiress Elizabeth Jaquelin. The property had been inherited by Elizabeth and technically belonged to her, but it took on the name of her husband, as was custom.

The Ambler house was built around 1750, but the human skeletons unearthed when the foundation was being dug suggest the property had been inhabited long before then. The mysterious human remains are believed to have come from a mass grave of the very first group of English settlers, perhaps those who died during the Starving Time during the winter of 1609–10. The remains might be from the mutilated corpses of the 350 settlers who died during the Indian Massacre of 1622, or they could be from settlers who managed to survive both those ordeals only to die from one of the many virulent diseases of the time. Despite the grisly discovery slowing down construction, the house was finally finished, and the Ambler family eagerly moved in. The eldest Ambler daughter, Lydia, soon fell in love with Alexander Maupin, a Patriot soldier with little money, property, education or pedigree. After a whirlwind courtship in which Lydia was completely and utterly swept off her feet, she married well beneath herself and proudly became Lydia Maupin.

There was little time to celebrate the union, though, because Alexander went off to fight in the Revolutionary War. Lydia remained at her parents' home, anxiously awaiting Alexander's return. As she waited day after day, she would venture down to the boardwalk that led from the house to a private pier overlooking the James River. She spent hours gazing out at the water, patiently waiting for her love to return to her.

Lydia and her family were forced to flee when Benedict Arnold set fire to their house in 1781, a particular shame because the war officially ended later that very same year. After the war, the Amblers returned to their island home

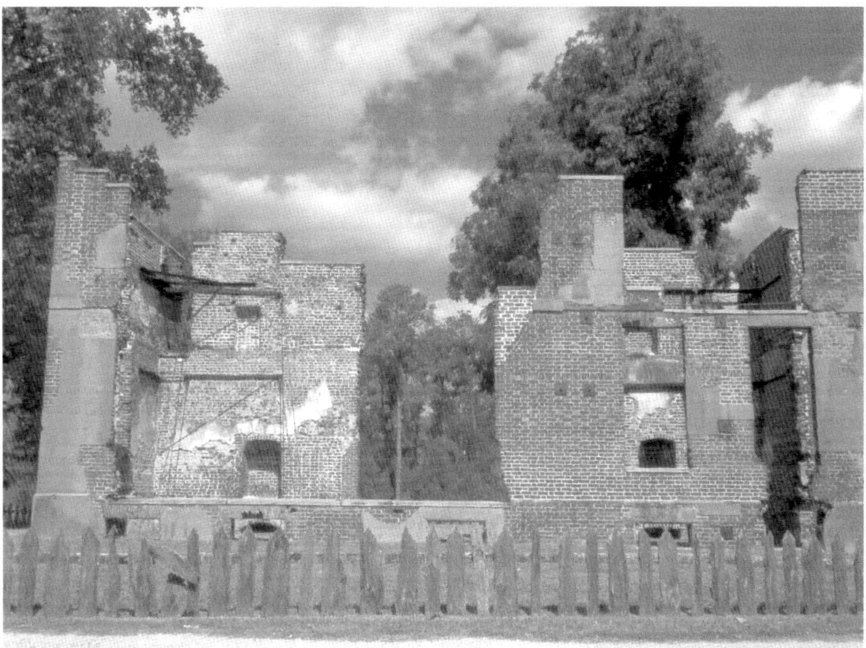

Ambler Mansion. *Photo by author.*

and set about rebuilding their damaged house. Lydia resumed her daily vigil on the pier, until one day, she grew tired of waiting and committed suicide by drowning herself, presumably by jumping off the pier she had spent so much time on. Alexander never returned home, and he is believed to have died in the war and been buried in an unmarked, hastily dug grave.

Lydia's father, Richard Ambler, died in 1766, and his estate was split between his three remaining sons. The property stayed in the Ambler family until David Bullock bought it, along with the Travis property, in 1831. The house was burned down during the Civil War but was quickly rebuilt. When it burned down again in 1895, it was left in ruins, which is how it looked when the National Parks Service acquired it in 1934.

In fact, most of the ghostly sightings have been reported by former Parks employees and tour guides. Supposedly, Lydia's spirit can be seen in the form of a flickering light dancing over the water where the pier once stood. This same flickering light has been seen wandering the property as well, perhaps still waiting for her long lost Alexander.

CARTER'S GROVE

We will leave this desolate spot and head to Colonial Williamsburg next, but along the way, we will need to stop at Carter's Grove, formerly called "Martin's Hundred" by the early English settlers. It is located six miles southeast of Williamsburg on Route 60 with a highway marker that reads:

> *Martin's Hundred: On both sides of this road and extending west was the plantation known as Martin's Hundred, originally 80,000 acres. Settled in 1619, this hundred sent delegates to the first legislative assembly in America, 1619. In the Indian uprising of 1622, 78 persons were slain here.*

The Virginia Company of London was the first venture to establish a settlement at this location, calling it Wolstenholme Towne; it was the hardest-hit settlement in the Indian Massacre of 1622. The original settlement was abandoned and had vanished by the time Robert Carter purchased the property in 1709. Robert Carter built the house on the property for Carter Burwell to live in, and the two Carter names give the plantation its current name, Carter's Grove.

Robert Carter III, grandson of Robert Carter, is perhaps best known for the emancipation of his five hundred slaves seventy-two years before the

Emancipation Proclamation of 1863. He thought slavery was "contrary to the true principles of religion and justice," and he set it up so his slaves continued being free for eight years after his death.

Robert Carter didn't share his grandson's beliefs, though, and after he finished the house at Martin's Hundred, he moved to Williamsburg to live with his family and slaves. Carter Burwell moved into the house in 1750, where he lived for six months before dying and leaving the property and house to his wife. The Burwell family was a typical slave-owning southern family of the antebellum period. They owned many slaves, including an entire African American family consisting of Jim; his wife, Betty; and their three children.

In 1765, the Burwells rented Jim out to work weekdays at the Governor's Palace, located eight miles away in Williamsburg. Jim was forced to walk the eight miles home to Carter's Grove every Saturday evening so he could spend Sunday with his family. Then, he was forced to walk the eight miles back to Governor's Palace every Sunday evening to be there in time for work on Monday morning. A few weeks into the arrangement, Jim returned home on a Saturday night to find his family had also been rented out. More than likely, they had been sold, and no one would tell him where they had gone. Jim never gave up hope and continued searching for his family until his death—and perhaps even beyond—but they were never reunited.

The plantation remained in the Burwell family until 1838, when it was already famous and visited by tourists who described it as "the most beautiful house in Virginia." Then, it passed through ten different owners over the next one hundred years. During this time, it was reported that the remains of three pirates were found buried beneath the cellar, probably part of Captain Guittar's crew, which you will meet in Chapter 7. In 1928, one of the last private owners made the very first alteration to the house—and quite a drastic one at that—when they added an entire third floor to it.

The Rockefeller Family Trust obtained the property in 1964 and passed it along to Colonial Williamsburg in 1967. Soon after, the Rockefeller Museum was opened, exhibiting artifacts that had been discovered at the original Wolstenholme Towne settlement. The slave quarters were reconstructed on the property and opened to tourists. It is around these same slave quarters where tourists have spotted Jim's wandering spirit. The property changed hands for the final time in 2008, when a private owner bought it for $15.3 million, and the contents were auctioned off a short time later. Unfortunately, it's no longer open to the public for tours, but it is clearly visible from Route 60, which we will continue on until coming to Williamsburg, our second point on the Historic Triangle and our next city to explore.

4

WILLIAMSBURG

Despite the crush of summer tourists, the colonial section of Williamsburg is a great place to visit, especially for curiosity seekers such as ourselves. Williamsburg is also a great place to live—just ask the singer Bruce Hornsby, who was born, raised and still lives there. Colonial Williamsburg is the second point of the Historic Triangle on our tour, and it was the capital of Virginia from 1699 until 1780. The second oldest college in America, the College of William and Mary, was established in Williamsburg in 1693, and the very first theater in America was erected there in 1716.

Originally, Williamsburg was called the Middle Plantation. It is still technically part of the Middle Peninsula area of Hampton Roads, but it is worthy of a chapter all to itself just for its historical value during the Civil War. Williamsburg is the place where Union and Confederate soldiers clashed in 1862, causing the rainwater in the streets to supposedly run red with blood as a result of the horrific skirmish.

Williamsburg was severely damaged during the Civil War, and it soon fell into a state of disrepair. Many homeowners never returned after the war, and the neglected buildings didn't recover until the Rockefellers began to restore them. John and Abbey Rockefeller enjoyed staying at Bassett Hall, after they had restored it, on their semiannual trips to Williamsburg. Bassett Hall is a rather nondescript white house built around 1760, with a small graveyard on the grounds. It was named after Burwell Bassett, Martha Washington's nephew. The Durfey family bought the house from Bassett, and they lived there during the Civil War. Their eldest daughter,

Margaret, fell in love with a wounded Confederate soldier while she was nursing him back to health. The soldier, John Lea, and Margaret were married after the Battle of Williamsburg, which took place on May 5, 1862, with approximately four thousand casualties. John's good friend George Custer of the Union army stood beside him as his best man. An inscription commemorating the event was etched into a ground-floor window of the Bassett house and is still visible today.

John and Margaret had six children; one died in infancy. They moved to West Virginia, where Margaret died in 1883. John remarried the following spring but died from blood poisoning eleven days later. He was buried beside his first wife in St. Albans, West Virginia.

A fire in 1930 destroyed much of Bassett Hall, but the Rockefellers soon restored it. Abbey Aldrich Rockefeller decorated the house in her favorite folk art style. Her husband, John, described the house as "a place to sit in silence and let the past speak to us." The house was bequeathed to Colonial Williamsburg in 1979 and was open to the public in 1980, still decorated in Mrs. Rockefeller's hand. The house was restored from 2000 to 2002, and then it was reopened to the public and remains so today.

The Rockefellers loved Williamsburg and were intent on restoring it to the grandeur of its pre–Civil War days. The Rockefeller restoration started funding the reconstruction of Colonial Williamsburg in 1926 and did not finish the project until 1940. A total of $52.6 million was spent to restore and rebuild approximately five hundred buildings in Colonial Williamsburg. The properties the Rockefellers couldn't buy outright were purchased with a life-right occupancy term—the owner was permitted to live in the building until death and then the property would pass to Rockefeller Restoration. One of the few properties they couldn't obtain was the Bruton Parish Church, part of which was (and still is) owned by the Bray family, who obstinately refused to part with it. Otherwise, the buildings were returned to their former antebellum glory, beginning with the Wren Building at the College of William and Mary, which is our first stop in Williamsburg.

College of William and Mary

James Blair founded the college, which was approved by King William III and Queen Mary II of England, in 1693. In 1691, Lieutenant Governor Frances Nicholson pledged funds for the college to be built but said funds hadn't materialized by 1697. In a letter to the Archbishop of Canterbury in 1697, Reverend Stephen Fouace of York Parish complained about Nicholson's negligence: "If we continue much longer under his Government we must...give over our endeavors to further what he hinders and uphold what he undermineth with all the craft he is master of."

Regardless of Nicholson's lack of contributions, the College of William and Mary was built, becoming the second oldest college in America, after Harvard. The first three original buildings were the Wren Building, flanked by the President's House and the Brafferton. Construction on the Wren Building began in 1695 and was completed in 1699, but it wasn't called the Wren Building until 1931 after rumors circulated that it had been designed by the famous architect Sir Christopher Wren. The chapel wing of the building was added in 1732. The Wren Building survived three major fires (in 1705, 1859 and in 1862), and it was rebuilt each time in a slightly different design.

Originally, the college was religiously affiliated with the Church of England and contained three different schools: pre-collegiate, philosophy and divinity for Anglican priests. Only after the Revolutionary War did it become a private school. Over the years, the college has been attended by several famous Americans—actress Glenn Close, Jon Stewart from TV's *The Daily Show*, James Monroe, John Tyler and Thomas Jefferson. Jefferson played the violin and cello and studied philosophy, mentored by George Wythe.

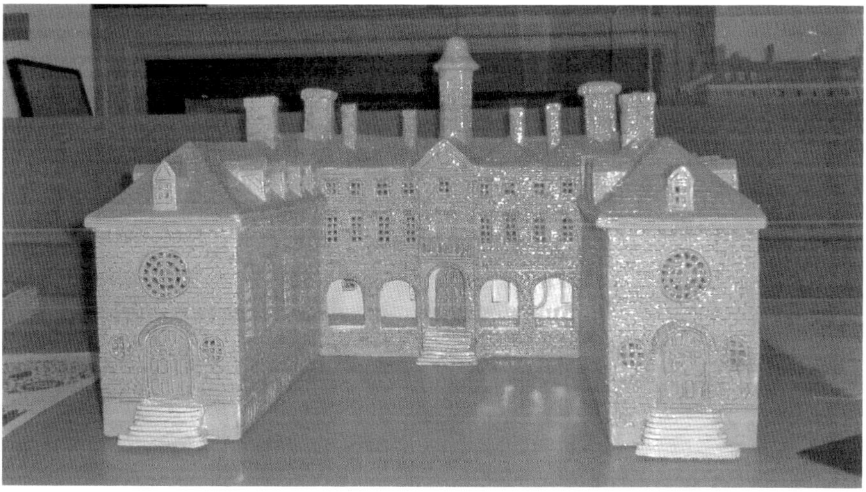

A replica of the Wren Building. *Photo by author.*

George Wythe (1726–1806) was born in Hampton but lived in Williamsburg as an adult. He replaced Peyton Randolph in 1761 on the board of visitors at the College of William and Mary, where he also taught students such as Thomas Jefferson and James Monroe. George Wythe was friends with George Washington and Lord Botetourt, and his signature appears on the Declaration of Independence because he opposed British rule, along with slavery. Wythe was the mayor of Williamsburg from 1768 until 1769, was elected to the vestry of Bruton Parish Church in 1767 and attended the second Virginia Convention, where he heard Patrick Henry's "Give Me Liberty or Give Me Death" speech. Despite all of his great deeds, he met with a tragic end. Within months of his sister's grandson—named George Wythe Sweeney—coming to live with him, Wythe died from arsenic poisoning. Days later, Sweeney attempted to cash checks signed by Wythe. Sweeney was arrested and charged with the murder of his namesake; Sweeney was acquitted but was found guilty of check forgery. George Wythe was buried at St. John's Episcopal Church in Richmond, although he probably would have preferred to have been laid to rest at the College of William and Mary beneath the Wren Chapel.

The chapel in the Wren Building contains numerous plaques and a gilt bust of the Royal Coat of Arms of the United Kingdom. Beneath the chapel floor lies a damp, musty crypt, unadorned save for a bit of graffiti, despite its not being open to the public. The crypt contains the remains of fewer than a dozen noteworthy Virginians. Sir John Randolph, a descendant of Pocahontas, and his wife, Lady Susanna, are buried together in one vault. His sons, John and Peyton, are also buried in the crypt. The ashes of Peyton Randolph and his wife, Elizabeth, were transferred to a lead coffin in 1929, along with the ashes of Bishop James Madison and possibly his wife. Lord Botetourt, the first governor of Virginia in 1768, is buried in the crypt, and a statue was erected outside of the Wren Building since it couldn't be placed over his body in the actual chapel where his relatives originally requested it stand. Chancellor Robert Nelson and former college president Thomas Dew are also buried in the crypt. It is believed that Judge Thomas Nelson is likewise buried in the crypt, but his exact burial place is unknown. Legend has it that a student drowned in a nearby lake and was buried in the vault, but no evidence has been found to prove this allegation.

All of the memorial tablets accompanying the deceased can be found on the walls of the Wren Chapel. Anything of value was removed from the crypt long ago because it had been broken into more than once. On one occasion, silver plating from Nelson's coffin was stolen and then later returned by a

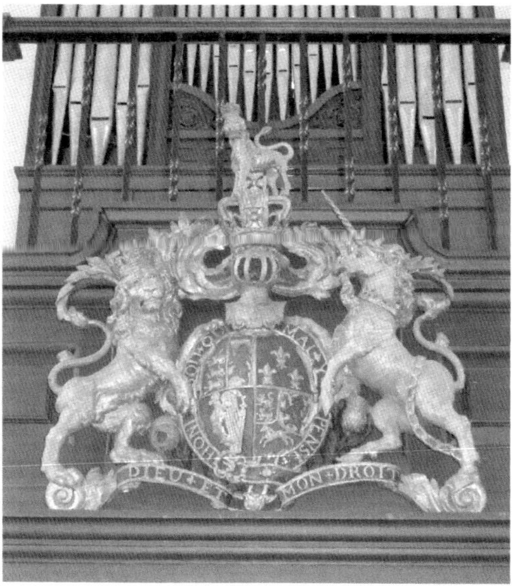

The Royal Coat of Arms in the Wren Chapel. *Photo by author.*

northern pawnbroker who had unwittingly bought it. On another occasion, Lord Botetourt's bones were strewn about, presumably as a joke, but rumors spread that the incident involved some sort of satanic ritual.

Now, only college personnel can access the crypt with a key and the alarm code. Without a guide, a person could get lost in the basement maze and completely miss the small door leading to the crypt. Once inside the vault, there are brick tombs on either side, and through a few small doorways resembling windows, the rest can be found. Directly next door are the steam tunnels, leading to the historic campus section and running beneath the Sunken Gardens. Over the years, numerous college students have illegally accessed the steam tunnels, having no idea they were only a few feet away from a crypt.

We will leave the college and venture on throughout Colonial Williamsburg, but the tunnels and vault at the College of William and Mary won't be the only ones we will encounter beneath the streets of Williamsburg. Our next stop will be at the Public Hospital; regardless of its name, the hospital is closed to the public but is visible from Francis Street. This two-story brick building with a dome-capped squat tower atop the roof is the second hospital to have been built in Williamsburg. The original hospital opened in 1773 at a different location under the official name "Public Hospital for Persons of Insane and Disordered Minds."

PUBLIC HOSPITAL

Public Hospital is the oldest mental institution in America and originally contained twenty-four barred cells with iron rings in the walls for restraining the patients and lice-infested straw bedding on the floor for them to sleep on. That was only the beginning of the severe mistreatment inflicted on the patients who had the misfortune of being housed there. Those poor individuals were also bled, plunged into ice baths and given electric shock therapy as part of their routine psychiatric treatments, not to mention the rancid scraps of food they were given, their chronic isolation and their cockroach and rodent companions.

Dr. John Galt joined the staff in 1841, incidentally around the time that conditions in the hospital began to improve. The patients were no longer shackled in their cells, were each given a proper bed and blanket and were fed three healthy meals a day. Plus, they were introduced to positive reinforcement for the first time in many of their lives, and part of their new therapy included social activities. Dr. Galt truly seemed to love his job, and he worked diligently at the Public Hospital for the next twenty years, constantly improving conditions for his patients.

When Union soldiers confiscated the hospital in 1862, they literally had to drag Dr. Galt out at gunpoint and threaten to shoot him if he returned. He was, however, permitted to remain living in his house on the hospital grounds, to which he retreated in a state of utter depression. The patients he had so lovingly cared for were neglected, and many died from the harsh new conditions. Other patients were released by the Union soldiers and allowed to roam the countryside, forcing them to inadequately fend for themselves and terrorize the residents who hadn't fled.

Two weeks after Dr. Galt was unceremoniously evicted from the hospital, he committed suicide by ingesting a large quantity of laudanum. He took so much that the blood vessels in his brain burst, and a pool of blood was found on the floor beneath his body. After his funeral, the house stood empty for a time before being inhabited by the Lee family, and Mrs. Lee wrote a letter about the memorable event that occurred on her first night in the house. Apparently, Mrs. Lee had vigorously scrubbed away the bloodstain left by Dr. Galt right before she went to bed. The next morning, a fresh bloodstain greeted her, and it wouldn't go away no matter what she treated it with or how hard she scrubbed. As if that weren't bad enough, the Lee children woke up terrified at night claiming to have seen a ghost at the foot of their bed, whose description uncannily matched Dr. Galt's appearance in life.

Following the Civil War, the hospital steadily declined and was renamed the Eastern Lunatic Asylum. It operated until an electrical fire destroyed it and several outbuildings in 1885. It is believed that 3 female patients died in the fire, and it is well documented that a total of 224 patients were left

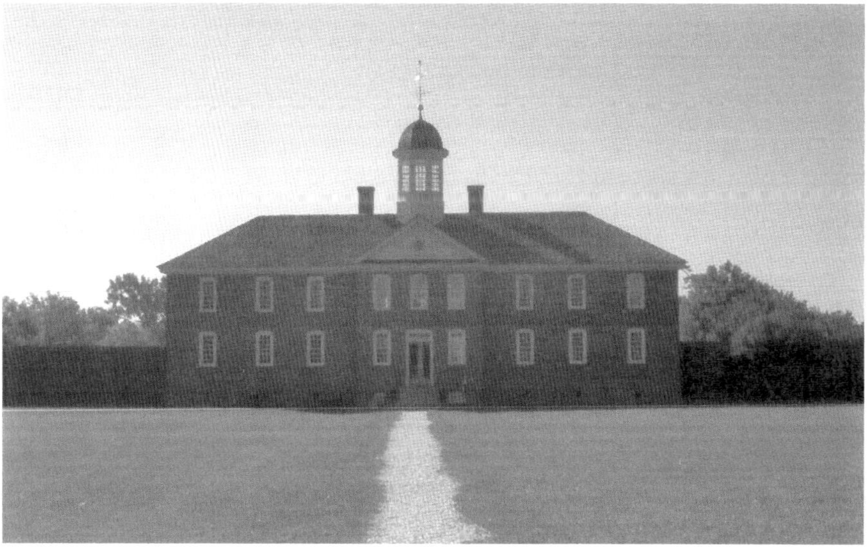

Public Hospital. *Photo by author.*

without a home. The hospital was rebuilt and renamed once again, to the Eastern State Hospital.

Colonial Williamsburg acquired the property in the 1960s, and the hospital was moved to a new location, with the land provided by the Rockefellers. Several years later, in 1972, an archaeological excavation took place on the original property, and the foundation of the first hospital was uncovered. The hospital was renamed again in 1985, to the Public Hospital, when it was reopened for a third time. It also had an exhibition on the first floor for tourists that included six cells, a diminished replica of the original hospital. Since then, however, it has been closed from the public's prying eyes. The house that once stood on the hospital grounds has long since been torn down, but the hospital itself remains intact and operational, though not at the same location as the original.

We will continue to stroll along until we come to Duke of Gloucester Street, where we will visit a few interesting buildings. The first is the Bruton Parish Church, one of the first Anglican churches built in America and home of the first American conspiracy theory. Bruton Parish Church is located at 331 West Duke of Gloucester Street, situated directly across the street from Martha Washington's childhood home. The original church was built in the 1600s, but the one you are seeing was built in 1715 and is still active today.

Bruton Parish Church

The earliest church records date back to April 18, 1674, with the Reverend Rolandus Jones presiding, who was an ancestor of David Roland Jones, Martha Washington's grandfather. Colonel John Page donated land for the church in 1678, and the church building was completed in 1683. Members have included Thomas Jefferson, Patrick Henry and George Washington, and extra services were held leading up to the Revolutionary War.

A cemetery surrounds the church, which we will investigate further in a moment, and two of Martha Washington's children with her first husband are buried next to the back wall of the church. Other notable people interred there include the Brays, who own part of the churchyard; Dr. Galt, the doctor at the Public Hospital; Colonel John Page; Reverend Jones; and Constitution signer John Blair. Some of the graves in the cemetery are unmarked and sunken, and others contain two or more bodies stacked atop one another. During the Civil War, the church was used as a hospital, and wounded soldiers were laid out on top of the cemetery plots as they awaited treatment.

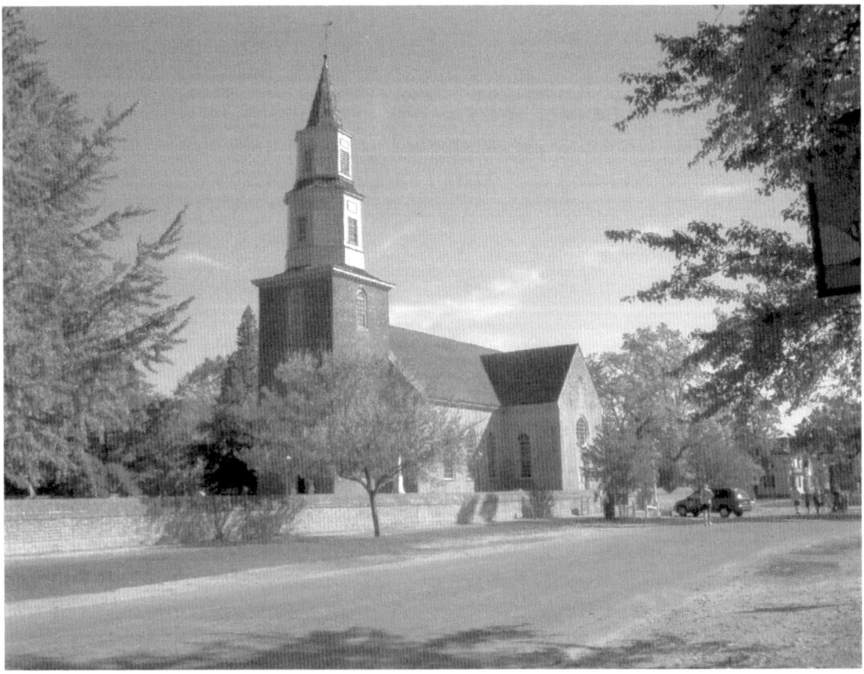

Bruton Parish Church. *Photo by author.*

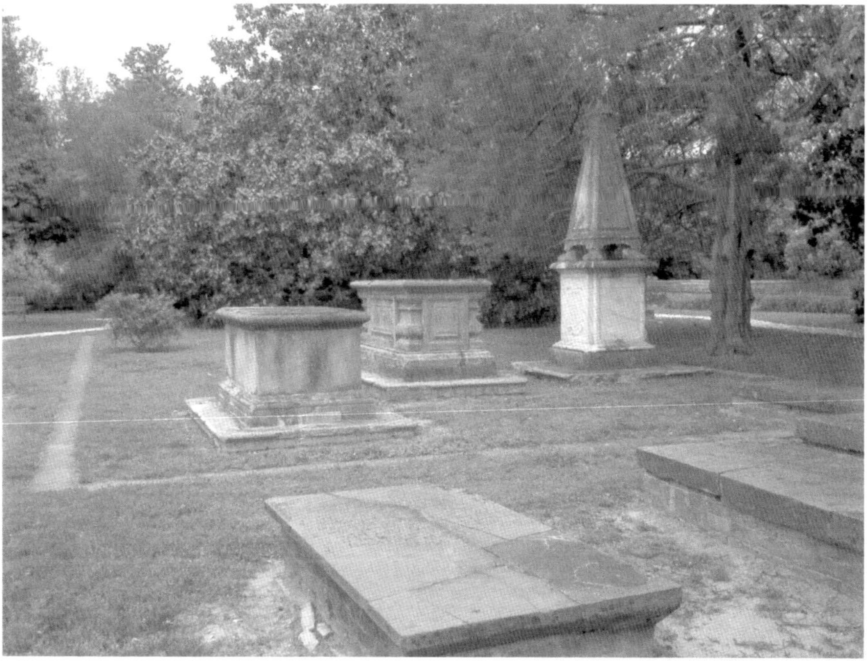

The cemetery at Bruton Parish Church. *Photo by author.*

The church only sustained minor damage during the Civil War, but it fell into a state of disrepair until the late 1800s. When the church was restored, previously unknown graves were found beneath the floor of the church. Further renovations took place throughout the early 1900s, mostly due to Reverend William A.R. Goodwin, who was also a part-time preacher at the Grace Episcopal Church in Yorktown, which we will visit in Chapter 6. To commemorate the 300th anniversary of Jamestown, in 1907, King Edward VII of England presented the church with a large Bible, and President Theodore Roosevelt donated a bronze lectern on which to display it, both of which are still in use today.

There is supposedly a mystery buried beneath the church and graveyard involving Sir Francis Bacon. Francis Bacon was born on January 22, 1561, in England. He was the youngest son of Nicholas and Anne Bacon. Bacon was knighted in 1603 and made the attorney general and lord keeper of the seal by King James I. Bacon went on to become the baron verulam in 1618, the lord chancellor of England and published several of his books, creating the concept of scientific methodology, quite unlike Aristotle's teachings. In 1621, Bacon became viscount of St. Albans, and soon thereafter, he was accused of taking a bribe, a crime he was eventually found guilty of committing. He was a self-made man; his father had died when he was only eighteen, and being the youngest, he had inherited nothing. He was completely penniless and thoroughly disgraced.

In March 1626, Bacon experimented with the effects of cold temperatures on meat, which involved stuffing snow inside a chicken carcass. Bacon caught a cold during said experiment, and the cold turned into bronchitis, which he died from on April 9, 1626. He was buried at the Parish Church of St. Albans in England.

Sir Francis Bacon's claim to fame is for allegedly being a member of a secret society that came to America under the guise of Freemasonry. Supposedly, Bacon translated the original version of the King James Bible but omitted certain passages and added others, thereby uniting all religious sects across the globe. Not only did Bacon rewrite the Bible, he is presumed to be the true author of William Shakespeare's works, the best cover-up in American history. According to legend, there are over one hundred vaults buried across America that contain evidence to verify this claim. One is thought to be located at the Bruton Parish Church, perhaps near a corner of the church beneath the cemetery.

Some conspiracy theorists believe that Captain John Smith transported Bacon's book to Jamestown in 1607, but the only contact between Smith and Bacon took place in 1618 when Smith asked Bacon to fund a colony in New England. Others think that it was Bartholomew Gosnold who brought Bacon's book to America because "Bartholomew" was a character in Bacon's *New Atlantis*. Rumors of this alleged vault and its priceless contents have persisted for so long that a search of the church took place for the first time in 1938 by Marie Bauer-Hall, funded by the Rockefeller Restoration. Later, she took part in an unauthorized search of the cemetery. This very public excavation took the better part of a day and uncovered the original 1683 foundations. Everyone who gathered around the excavation thought the foundation would be the only thing there, but twenty feet down, a large box was unearthed. The box was approximately ten square feet and was considered by eyewitnesses to be too large for a simple coffin.

It was a well-publicized event, and a crowd of curious photographers and reporters had been waiting outside the church fence all day. But when the box was uncovered with a resounding "clang," the crowd scrambled to the top of the fence to snap pictures. Church officials feared a riot would ensue, so they ceased the excavation without any explanation. Later, those same church officials stated that the great discovery had only been a grave,

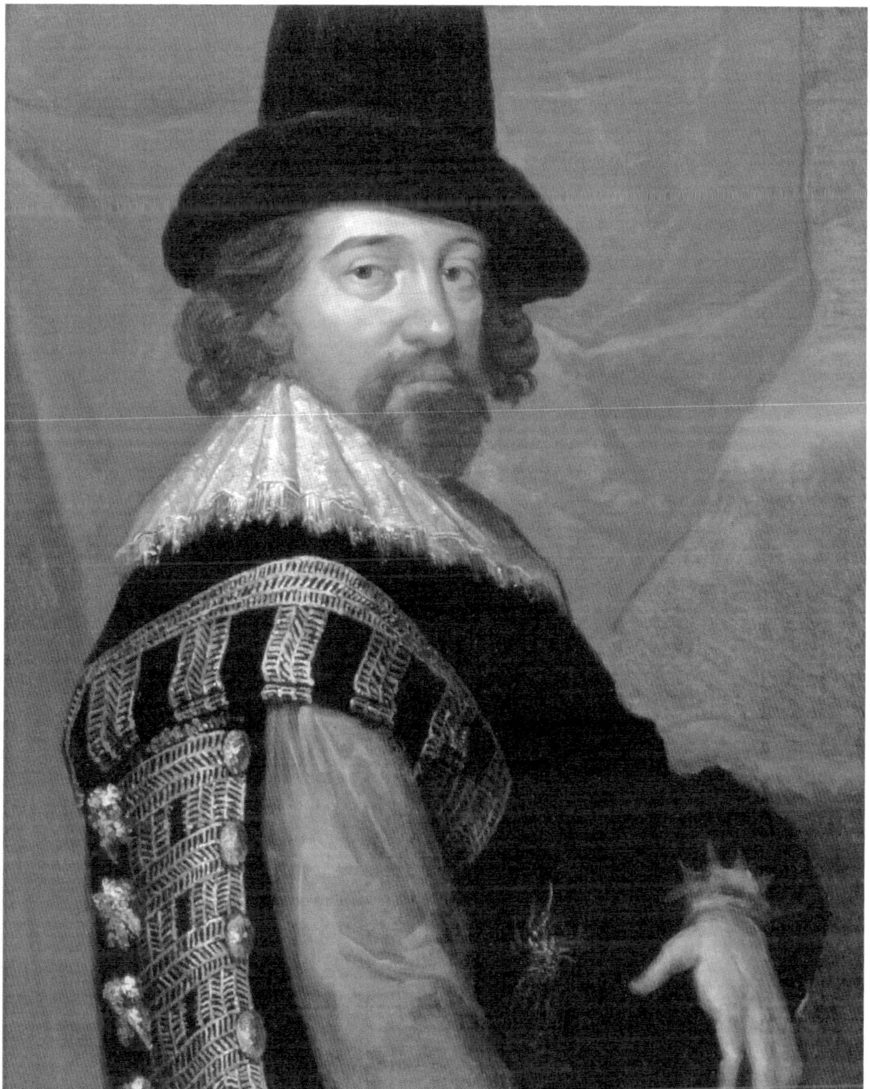

Sir Francis Bacon. *Photo courtesy of Wikimedia Commons.*

and they had halted the proceedings because they had no desire to disturb someone's final resting place.

Finally in 1985, more authorized scientific testing was permitted by the church officials. The results of the test were inconclusive but suggested something was indeed below the surface of the cemetery other than graves.

In 1991, two separate illegal digs took place at the cemetery, and one of them put a hole in the church wall before being halted by officials. Conspiracy theorist Marsha Middleton participated in both digs. She still believed something was under the churchyard, and her diligence paid off a year later when another authorized excavation took place, but it turned up nothing more than the grave of a child and some really old dirt. In 1992, another authorized excavation took place, this time by professional archaeologists, who found no evidence of tunnels or vaults.

Yet rumors of the vault still persist. Some theorize that church officials had plenty of time to remove the damning evidence in 1938, not wanting to ruin Shakespeare's reputation and change history in one fell swoop. Many wonder what sort of documents would be contained in such a large box. After four centuries, it still remains a mystery, and it might for four hundred more—if it is ever discovered. To locate the vault, you will need to read Bacon's novel *New Atlantis*, published a year after his death in 1627. Supposedly, there are encrypted messages in the book that correspond to clues around the Bruton Parish Church. These clues are encoded in certain headstones around the cemetery and point the way to the vault.

The first of these headstones is found inside the church and belongs to Nathaniel Bacon the Elder. His epitaph reads:

> *Here lyeth interred ye body of Nathaniel Bacon Esq whose descent was from the Ancient House of ye Bacons (one of which was Chancellor Bacon and Lord Verulam) who was auditor of Virginia and President of ye Honourable Councell of State and Commander in Chief for the County of York having been of the Councell for above forty years and having always discharged ye offices in which he served with fidelity and loyalty to his Prince who disparted this life ye 16 March 1692 in ye 73 years of his age.*

Francis Bacon was related to Nathaniel Bacon the Elder, who came to America in 1650; one of the original founders of Jamestown, Bartholomew Gosnold; and Nathaniel Bacon the Rebel, whom you will learn more about in Chapter 5. Nathaniel Bacon the Elder was the son of Francis Bacon's cousin Reverend James Bacon. Bartholomew Gosnold was the grandson of Francis Bacon's cousin George Bacon, and Nathaniel Bacon the Rebel was the son of Francis Bacon's cousin Thomas Bacon.

But before you start digging beneath Nathaniel Bacon's stone (or anywhere else), you need to get permission from church officials, which is impossible to obtain because they deny the vault's existence; they are backed by archaeologist Andrew Edwards, who excavated the churchyard in 1992. While it is a little-known secret that there are various crypts, vaults and tunnels under the ground of Williamsburg, a few have been unearthed from time to time.

A chest filled with pre–Revolutionary War papers—written by George Washington, Thomas Jefferson, John Page and others—was found in a vault beneath the Robert Carter House. This chest was damaged during the Civil War and has subsequently been lost. The Robert Carter House was bought in 1761 by Charles Carter, son of Robert Carter from the Carter's Grove section in Chapter 3. The house changed hands a few times, coming into the possession of Dr. Kenneth McKenzie, a surgeon who worked with Dr. Galt at the Public Hospital. Dr. McKenzie opened an apothecary next door, which is still open for business, connected to the house by a verandah. Eventually, the house was bought by the Robert Saunders family, who were forced to flee from the invading Union army during the Civil War, leaving supper dishes on the table in their haste. Before Federal Marshal Major Wheeling was posted in it, Union soldiers plundered the abandoned house, and it is believed that this is when the chest was found and damaged.

Officials from the City of Williamsburg vehemently deny the existence of any underground tunnels, other than drainage ditches, yet some of these "nonexistent" tunnels have been found, their original entrances sealed during the Rockefeller Restoration project in the 1930s. There are also archaeological reports in which vaults were inventoried beneath the Robert Carter House, the James Geddy House and Governor's Palace. Tunnels have been located running between Governor's Palace and the Peyton Randolph House, and an unexplored tunnel was found meandering away from the Benjamin Waller House. All of these vaults and tunnels are listed in the *1953–56 Archaeological Report of Williamsburg*. With such obvious proof, it is hard to deny their existence.

Governor's Palace

One of the buildings mentioned, Governor's Palace, is our next stop in Williamsburg and is located near the Bruton Parish Church at the end of the Palace Green. Governor's Palace is a three-story brick building with dormer windows peering out of the shingled roof, which has a towering steeple atop it. It is one of the two largest buildings in Colonial Williamsburg. It was first dubbed a "palace" during its construction in 1714. The name stuck, and when it was finished in 1722, it was given the official name of Governor's Palace. Seven royal governors (meaning they had a title in their name, such as "Lord") have called the palace home over the years, along with Patrick Henry, Thomas Jefferson and most of the governors of Virginia ever since its completion.

Half a century after it was completed, the palace was used as a hospital for wounded Patriot soldiers during the Revolutionary War. Unfortunately, it burned to the ground on December 22, 1781, killing a bedridden patient who was trapped inside. The palace remained in a state of ruins, and in 1835, Sir Charles Augustus Murray visited the site and described it: "The centre of the palace where the governor resided has long since fallen down, and even the traces of its ruins are no more to be seen."

It continued to deteriorate and was further damaged in 1862 when Union soldiers pulled down parts of the remaining structure because they needed the bricks to build chimneys at nearby Fort Magruder. After the Civil War ended, the College of William and Mary acquired the site. Then, in 1928, Colonial Williamsburg purchased the property, along with the two school buildings on the grounds.

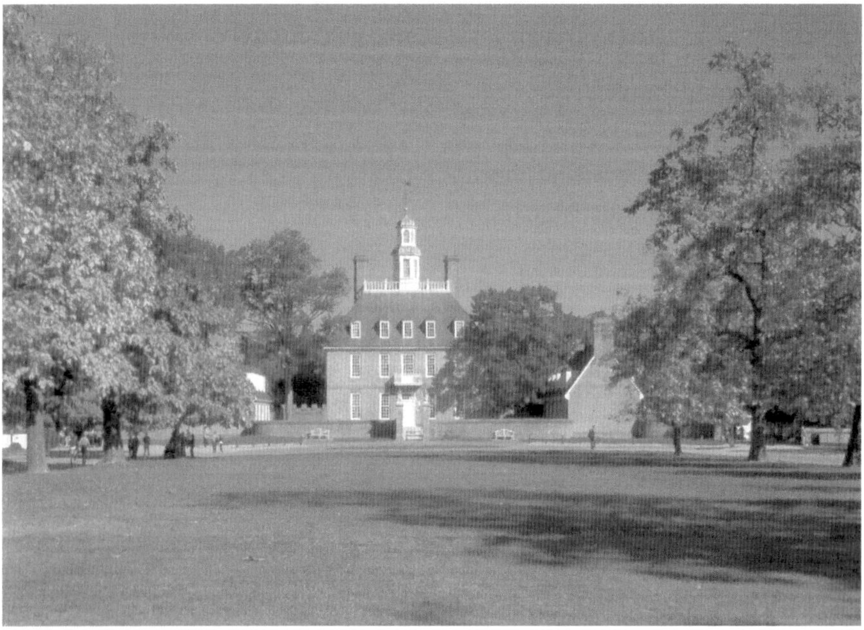

Governor's Palace from the Palace Green. *Photo by author.*

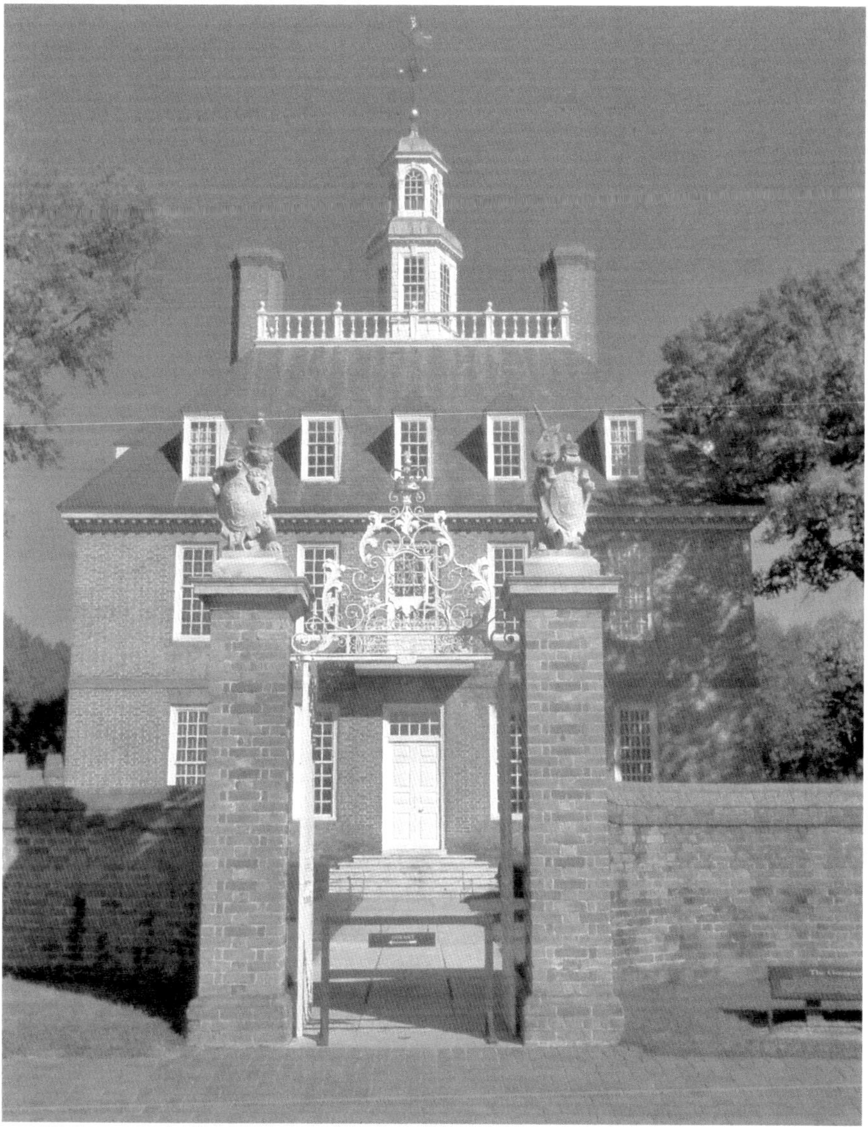

Governor's Palace. *Photo by author.*

The palace was completely rebuilt in the 1930s, with drawings and engravings of the original building used as blueprints to re-create it. An excavation of the site took place prior to construction, and a total of 158 human remains were discovered. Scattered among the bones were hundreds of musket balls with teeth marks embedded in them. Of the remains, 156 skeletons were Revolutionary War soldiers who had bitten down on the musket balls during surgery due to lack of pain-killing drugs. The remaining 2 skeletons were female and were probably nurses who died from one of the ravaging diseases of the time.

More updating to the palace took place in 1981 when an original 1770 inventory of the estate was used to replicate the interior furnishings. Yet even more renovations took place in 2006 for the new governor, Governor Dunmore, who replaced the former bachelor governor; Governor Dunmore had revisions made to accommodate his wife and their six children. No further construction has been done to the palace since then, and it stands in stately grandeur at the end of Palace Green, much like it did after it was first built.

Peyton Randolph House

About halfway down the Palace Green you will find Nicholson Street, on which our next stop is located, the Peyton Randolph House on the corner of North England and Nicholson Streets. Sir John Randolph, the first Virginian to be knighted by England's crown, built the original house in 1715. Soon he and his wife, Susanna, moved in. They eventually passed the house on to their twenty-four-year-old son, Peyton Randolph, who was Pocahontas's great-great-great-great-grandson. Peyton Randolph was also the grand master of the first Freemasons Lodge in Virginia. Peyton married Elizabeth "Betty" Harrison, and when he died in 1775, he left the house to her. He is buried beneath the Wren Chapel at the College of William and Mary.

During the Randolphs' possession of the house, many famous historical figures stayed there, including George Washington and General Rochambeau, prior to the Revolutionary War. Washington is a well-known figure, but Rochambeau is a bit more obscure, though he had a great influence on American culture all the same. General Rochambeau's complete name with title was Jean Baptiste, Comte de Rochambeau (Count of Rochambeau), and he was a French general in the U.S. Army during the Revolutionary War. General Rochambeau is best known for making decisions during combat by using a hand game, and roshambo, or "rock-paper-scissors," has been around ever since. Perhaps the general even showed the game to Peyton as a guest in his house, though it is doubtful Peyton's wife, Betty, would have approved of such frivolous acts as hand games.

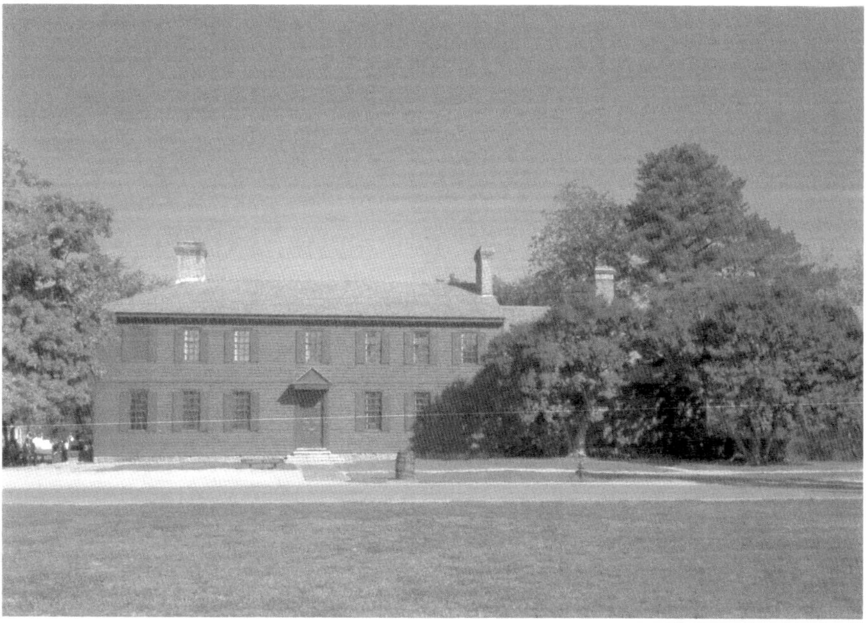

The Peyton Randolph House. *Photo by author.*

Betty was an imposing figure and a cruel master to her slaves, with her personal maid, Eve, bearing the brunt of Betty's cruelty on numerous occasions. Betty's treatment was so severe that Eve ran away twice only to be captured, returned and whipped as punishment. Betty eventually died, but her cruelty, which seemed to know no bounds, didn't stop with her death. From beyond the grave, Betty reached out and snatched away everything from Eve but her very life.

Betty's will stated that Eve should be sold, and two new slaves should be bought to replace her. This news was bad enough but was doubly heartbreaking because Eve was now a mother. Eve had to be physically pulled out of her son's arms when her family was split up. She was thrown facedown across the back of a horse with her wrists tied to her ankles under the horse's belly. As Eve was hauled away, she loudly cursed anyone who should ever dare to live in the house.

Peyton and Betty had no children, and after Betty's death in 1783, the house was auctioned off. The Peachy family bought the house in 1824 and were the first residents to believe Eve's curse was still in effect. One of the Peachy children died from falling out of a tree, and the rest of the Peachy children subsequently died from various diseases. One Peachy family member committed suicide, and a boarder staying in the house died from tuberculosis. To date, twenty-three deaths have been linked to the Peyton Randolph House.

Colonial Williamsburg acquired the house and began to restore it in 1938, with more restoration taking place in 1967 and 1977, this time to the outbuildings. Former employees, tour guides and guests of the house have seen the ghost of a woman inside. A workman once reported hearing strange voices in the empty basement as he repaired the furnace. But Eve's curse seems to have run its course, with no deaths in decades, so you shouldn't feel threatened or frightened.

PUBLIC RECORDS OFFICE

We will leave this cursed house and head to the site of another tragic tale at the Public Records Office, located on Duke of Gloucester Street. The Public Records Office, otherwise known as the Secretary's Office, is the oldest archival building in America. This squat, fireproof, brick building was constructed in 1748 to house important records and legal documents after the capital building burned down, destroying many records.

It was a court secretary who petitioned a bill to the House of Burgesses to build the office. The bill was passed, and the office was built to very precise specifications: brick walls inside and out, a stone floor and the only flammable objects were the chair rails mounted along the walls. Four fireplaces were installed to help preserve the books from humidity, or so the early colonists thought; the fireplaces were actually built to prevent downdrafts from throwing sparks into the room.

Over the years, the Public Records Office has been used as a house, a clerk of court's office, part of the Female Academy, an elementary school and even a hide-out for fleeing Confederate soldiers during the Civil War. Confederate soldiers were hiding in the building near the end of the war when Union soldiers surrounded it, trapping the hapless Confederates inside. The Union soldiers proceeded to open fire on the building before bursting inside and capturing the few soldiers who remained alive. Despite being fireproof, it was burned almost to the ground. After the war it was rebuilt, with musket ball holes still pocking portions of the few original walls left intact.

David Roland Jones (a descendant of Reverend Jones of the Bruton Parish Church), his wife and their seven daughters bought the building soon after the war was over and used it as their home. Jones was a strict, devoutly religious man who never let his daughters out of his sight; they certainly never left the house or got married. The Jones family rarely left the property at all, choosing instead to have supplies delivered to their house on a semiweekly basis.

One of the Jones daughters, Edna, fell in love with the delivery man, John Mince. Edna was unsure how John felt about her, but she knew her father would never approve, so she kept her feelings hidden for several long months. Unbeknownst to Edna, John had developed feelings for her, too, and one day during his delivery, he professed his love to her. At first, Edna was elated but quickly sobered when she realized her father would not be pleased, and the couple made hasty plans to run away together that very night. They agreed to meet at the Palace Green after dark, and John stole a quick kiss from Edna before he resumed his delivery route.

Edna impatiently waited for darkness to descend, and when she saw the first star in the sky, she sneaked out of her home. She hurried down the sidewalk, barely looked before she stepped out to cross the street and was struck by a passing horse-drawn carriage. Edna died on the street within eyesight of the Palace Green and her love. Edna's father buried her in the family cemetery, located in the yard of their home.

Coke-Garrett House

Our next stop, the Coke-Garrett House, is our last in Williamsburg. It is located at the corner of Nicholson and Waller Streets, next to what used to be the Public Gaol, or public jail. Waller Street becomes Capitol Landing Road after it crosses Lafayette Street, and Capitol Landing Road used to be called Hangman's Road because prisoners were forced to ride in a wagon atop their own coffins to the scaffolding at the Public Gaol. The Coke-Garrett House is a sprawling white plantation home and has been owned by a baker, a barber, a doctor, a diplomat and three former college presidents over the years, but it got half of its name from the original owner, John Coke.

John Coke was an accomplished silver- and goldsmith, and he owned the property in the mid-1700s. When he owned the property, there were at least two buildings on it—his house, which was built around 1755, and a tavern, which doubled as an inn. Unfortunately, Coke died in 1767, leaving his property and establishment to his wife, Sarah. She feared the locals would think the tavern had closed, so she took out an announcement in the paper: "Sarah Coke begs leave to inform those Gentlemen who were so kind as to favour her deceased husband with their custom that they may depend on receiving the same entertainment as formerly."

So the tavern remained open, and Sarah leased it to another tavern owner two years later, who offered "genteel lodgings and good entertainment." Sarah, savvy businesswoman that she was, even leased out her home during the Revolutionary War for the Patriot soldiers to use as an army barracks. The house was severely damaged, but the Virginia House of Delegates reimbursed her after the war was over.

In 1780, the house was abandoned after the capital of Virginia moved from Williamsburg to Richmond. Thirty years later, Williamsburg was inhabited once again, and the Garrett family bought the house in 1810 and returned it to its former glory. The house remained in the Garrett family throughout the Civil War when Dr. Robert Garrett lived there, the person from whom the second half of the name comes. Wounded soldiers were constantly being dumped on Dr. Garrett's front lawn during the Civil War, and his primary job consisted of amputating partially severed limbs. He performed countless surgeries. In fact, the amputated arms and legs lined his office walls up to the windowsills, and "amputation pits" had to be dug in his backyard. The amputees who died—and many of them did—were wrapped in the sheets they had died on and were buried in mass graves. Not all of the soldiers died from amputations; even more of them died from gaping chest wounds. According to legend, a Confederate soldier was helping his wounded comrade to safety when he was shot in the chest. Ironically, the wounded friend lived, but his rescuer died on the grounds of the Coke-Garrett House.

The grounds were professionally excavated in the 1950s, and broken pieces of pottery were mostly all that was found. A few other items uncovered were directly related to the first owner, John Coke, including jagged shards of wineglasses and fragments of precious metals. The only evidence of the next owner, Dr. Robert Garrett, is in the building's name.

We will leave Williamsburg now and its colonial houses, old cemeteries and secrets in order to travel to Gloucester. There, we will see more colonial houses and old cemeteries and learn about other curiosities and secrets, including betrayal, entrapment and a secret passageway. Many of the upcoming sites are well off the beaten path and virtually unknown to tourists, so we will be venturing into new territory.

5
GLOUCESTER

Westover Plantation

The next leg of our journey will take us through the Middle Peninsula to Gloucester, but we will need to make a pit stop along the way. The Westover Plantation is located on Route 5, in Charles City County, and is a gorgeous example of fine, Georgian-style architecture. The house boasts a steep roof with dormer windows poking out of it and towering chimneys on either end. The vast front lawn of lush, emerald grass gently slopes down to a small, private beach on the north bank of the James River. It also has at least one secret passageway that leads under the house to the James River and was built as an escape route from random Indian attacks.

William Byrd II was secretary for the Virginia Colony and was also the founder of the city of Richmond, but the Westover Plantation, which he built in 1726, was his crowning glory. The prosperous Byrds enjoyed hosting lavish parties in their elegant home, and everyone for miles around vied for invitations. The Byrds' daughter, Evelyn, missed out on the parties, though, because she was sent to England for a formal education at the tender age of ten. She remained abroad for years until she had become a breathtakingly beautiful young lady.

After her schooling was finished, Evelyn traveled to London, intending to board a ship bound for America within days. Instead, she met and fell in love with Charles Morduant, and they sent a letter to her father in Virginia asking for his permission for the couple to wed. Morduant was a Catholic, and Evelyn had been raised as a Protestant, so it was no surprise when her father returned a letter forbidding her from marrying the young man who had captured her heart. What did surprise Evelyn was the fact that her father threatened to disown her if she disobeyed his wishes and married Morduant.

Evelyn, heartbroken and humiliated, returned to her father's plantation in Virginia, never to see Morduant again. She was a bitter young woman when she arrived at Westover Plantation, and she became a recluse who refused to see any of the eligible Protestant bachelors her father attempted to thrust on her. She died in 1737, bereft and alone, at the age of twenty-nine. A perpetual cloud of doom seemed to hang over the house after Evelyn's passing, and a series of strange events occurred. Evelyn's sister-in-law, Elizabeth Hill Carter Byrd, was crushed to death in the attic of the house when a chest of drawers toppled over on her. Six months later, William Byrd III gambled away the family fortune and committed suicide as a result.

The house left the Byrd family in 1817, and the West family moved in. Henry West was the fourth Lord of Delaware (De La Warr) and the son of Governor Thomas West. In 1899, Clarise Ramsey, a descendant of the Byrds, bought the house. The last Byrd descendant moved out of the house in 1921, and the Crane family moved in, and it remains in the Crane family today. The Cranes generously keep the grounds and gardens open to the public, although the house itself is closed. Be sure to take notice of the towering tulip poplars in the yard, as they are close to two hundred years old. And don't miss the tomb located on the grounds that contains the remains of William Byrd I. He originally bought the property in 1688, was buried there in 1744 and was known as the "Black Swan of Westover."

Starvation Road

We'll leave the Black Swan to his peaceful slumber and continue traveling northeast to Starvation Road in Signpine, the northwestern portion of Gloucester near Adams Creek. The road is in the middle of nowhere, surrounded by fields and only its name provides a clue as to its historical relevance. During the Civil War, a group of Confederate Rebels had their headquarters in a swamp near present-day Starvation Road. These Rebels staged several successful raids against the Union soldiers who were stationed in Gloucester. Finally the Union troops found out the exact location of the Rebels' headquarters and quickly surrounded it, leaving only a single way in and out unprotected. Instead of a full-force attack, the Union troops decided to wait out the Rebels and then pick them off one by one as they attempted to retreat down the single unguarded path. However, the Confederates chose to end their lives on their own terms, and they slowly starved to death during the ensuing winter…leaving only the name of the path to survive, now called Starvation Road.

GLOUCESTER COURTHOUSE

We'll need to get on Route 17 and head southeast to our next destination, the Gloucester Courthouse in the downtown area. Located inside the courthouse is a site of historical importance. There is a plaque commemorating Nathaniel Bacon and the first revolution to take place in America, in 1676.

Nathaniel Bacon the Rebel was born in England in 1647, and he married Elizabeth Duke in 1674. Her father promptly disowned her, so the couple traveled to Virginia, where they resided about twenty miles southeast of the James River waterfalls. Governor Berkeley was the chief official in Virginia. There had been no new elections in years, and there were heavy taxes in effect, especially on the poorer classes. Plus, Berkeley was sympathetic toward the Indians, which was a big strike against him in Bacon's eyes because his overseer at Bacon's Quarter had been killed by Indians.

Bacon led five hundred men to ask Governor Berkeley's permission to march against the Indians; Berkeley refused and declared that Bacon's gang was illegal. All but sixty of Bacon's men deserted him and his cause; the remaining sixty who stuck by his side attacked the Indians a few times before they disbanded. Bacon was then arrested on his way to Jamestown, where he confessed and apologized to Governor Berkeley, his relative Nathaniel Bacon the Elder and several other council members. Then, Bacon took his newly appointed seat on the council of state.

But rumors began to circulate that Berkeley planned to kill Bacon, so he fled. Six hundred disgruntled colonists rallied around Bacon, and they all marched to Jamestown, quickly taking it over. Some temporarily quartered at Bacon's Castle in nearby Surry, forcing its current resident, Major Arthur Allen, to flee.

Bacon's Castle, located at 465 Bacon's Castle Trail, was built in 1665 by Arthur Allen, and the cruciform brick building with triple chimneys on each end of the original structure looks much like it did then. Upon Arthur Allen's

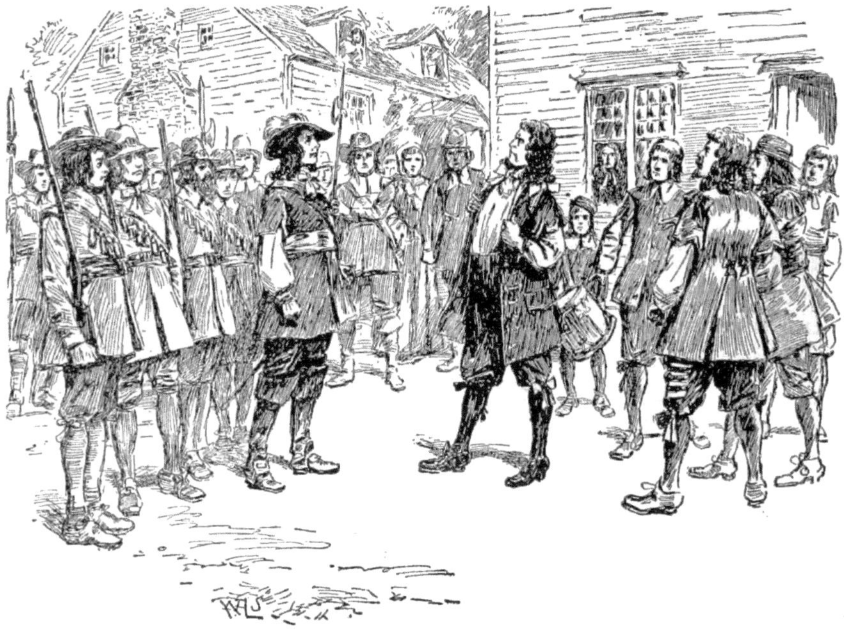

Bacon's Rebellion. *Photo courtesy of Wikimedia Commons.*

death, the house was passed to his son Major Arthur Allen. During the 1700s, Arthur Allen III and his wife, Elizabeth Bray (a member of the Bray family from the Bruton Parish Church section in Chapter 4), lived in the house. After that, the house passed through many hands until the Association for the Preservation of Virginia Antiquities (APVA) purchased it in 1972. The APVA restored the house and used the Allens' inventories from the 1700s to re-create the furnishings before opening it to the public for tours in 1983.

While Nathaniel Bacon and his rebels stayed there, a couple more Indian attacks against the English occurred, resulting in the death of 8 colonists. Bacon and his raiders headed off for justice, leading several attacks against the Indians during the spring of 1676. By summer, only 140 raiders remained by Bacon's side, the rest having deserted him. These remaining die-hards returned to Jamestown again to besiege it, and they captured Berkeley's wife and held her hostage. Later, Berkeley reported that Bacon had burned down five of his houses, along with twenty others and the church.

Bacon and his rebels traveled to present-day Williamsburg, where they issued a declaration of the people in which they made numerous complaints and accusations. The Declaration in the Name of the People of Virginia claimed that Governor Berkeley had heavily taxed the commoners in order to advance his favorites "and other sinister ends" but without any "visible effects in any measure" (during his time as governor, the colony hadn't advanced). The declaration accused Governor Berkeley of raising his "scandalous and ignorant favorites" to positions of justice. Furthermore, Governor Berkeley was accused of monopolizing the beaver fur trade.

The second part of the declaration concerned Berkeley's dealings with the Indians; he was accused of betraying the colonists to the "barbarous heathen" and turning the army of the English (Bacon and his rebels) back from the Indians. This caused the Indians to become "a very formidable enemy," "who might at first with ease have been destroyed."

The declaration went on to demand that Berkeley and his men should surrender or the colonists who hid them would be considered traitors, and their property would be confiscated. As a union, Bacon and his men would defend themselves against the common enemy: "Let not the faults of the guilty be the reproach of the innocent." Nor should the crimes of their oppressors divide them. Berkeley and his twenty "wicked and pernicious councilors" who were listed should be considered traitors, seized and then secured at Middle Plantation. The declaration was signed by "Nathaniel Bacon, General by consent of the people."

After issuing the declaration, Bacon and his rebels traveled to Gloucester, where they took over the home of the speaker of the House of Burgesses. Shortly thereafter, Nathaniel Bacon the Rebel fell ill and died there, his body hidden by his fellow raiders.

The marble plaque in the Gloucester Courthouse commemorates him and his efforts. It reads:

> *In memoriam, Nathaniel Bacon the Younger. General and member of the Governor's Council. Born in Suffolk, England—1630–40* [sic]*—died in this county in 1676. Originator of his so-called Rebellion, whose influence in the foundation of the spirit of Americanism is immeasurable—the Washington of his day, popular and patriotic, whose magnanimity strongly contrasted with Berkeley's malignity. A soldier, a statesman, a St. Gloucester, who honors the noble dead, and cherishes the memory of kingly men and in whose soil the body of Bacon is said to sleep, erects this monument to the great patriot, by the authority of the Circuit Court, through the generosity of his friends.*

ROSEWELL RUINS

We'll continue traveling southeast on Route 17 to a place that was once considered one of the most beautiful antebellum mansions but now stands in ruins. The Rosewell Ruins in Gloucester is located at 5113 Old Rosewell Lane and is surrounded by cornfields and woods at the end of a long, dirt lane. Only 2,500 tourists legitimately visit the site each year, but due to its remote location, the local youths frequently trespass, as the graffiti carved high into the interior walls and the number of unbroken beer bottles in the bottom of the ice pit can attest. Now is a good time to apply insect repellent and keep an eye out for snakes because this site is well populated by both pests.

The property was always called Rosewell Plantation because of a nearby stream called Rosewell Creek, but some longtime locals still refer to it as the Page Plantation. The prominent Page family, one of the first families in Virginia, acquired the property in the mid-1600s. The square, three-story brick house with chimneys at each corner was built between 1725 and 1738, and it remained in the Page family for the next one hundred years.

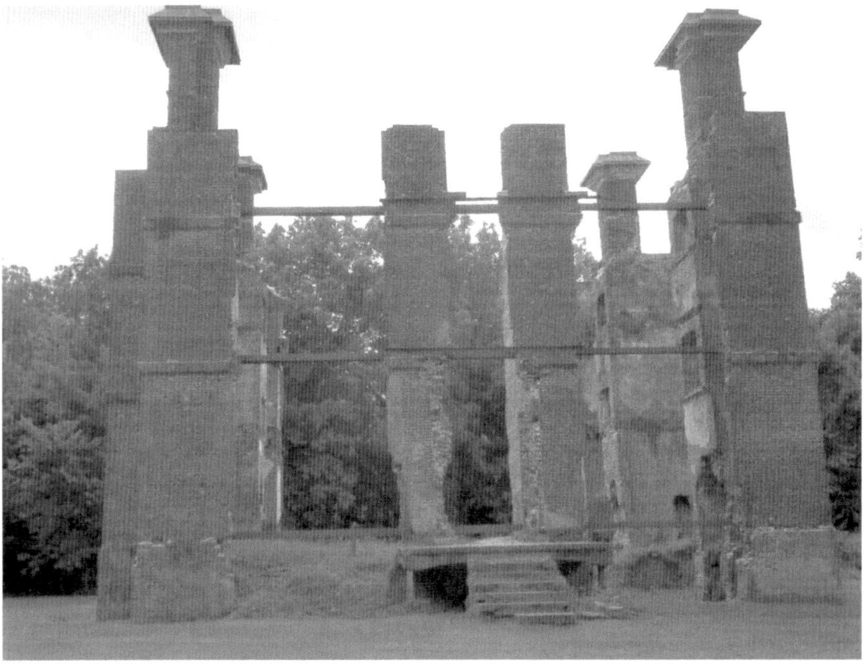

The Rosewell Ruins. *Photo by author.*

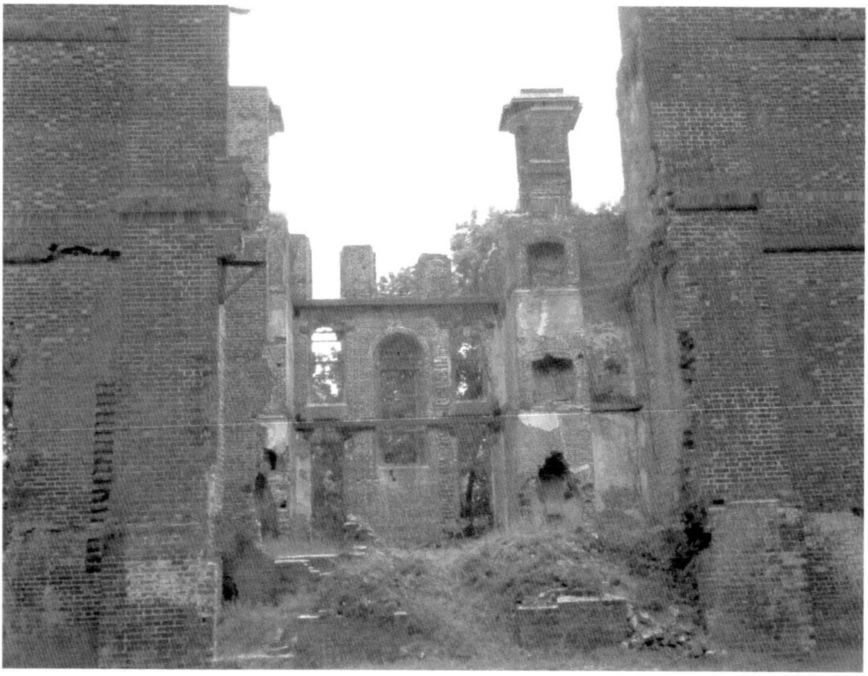

The Blue Room was located on the second floor, left foreground of photo. *Photo by author.*

Regardless of who owned it, the most sensational balls were held in the house until well after the Civil War.

John Page was born in 1744 and graduated from the College of William and Mary alongside Thomas Jefferson in 1763. The two men were close friends, and Jefferson frequently visited Page at his plantation. In fact, Thomas Jefferson even wrote the rough draft of the Declaration of Independence in the Blue Room of the Rosewell mansion while staying there as Page's guest. The Blue Room was located on the northwest corner of the second floor.

John Page became a colonel during the Revolutionary War and even supplied provisions for the Patriot troops out of his own deep pockets. Page went on to become a congressman in 1789 and retained the prestigious position on and off until 1797. In 1790, a much younger Scottish poet—Margaret Lowther, whom Page nicknamed "Peggy"—caught his eye.

Although Page had a couple of strikes against him, being a middle-aged widower and all, Margaret hesitantly allowed him to escort her to a party. She accidentally left one of her gloves in his carriage at the end of the evening, and Page returned it to her along with a note: "Taking 'G' from

'Glove' leaves Love, Tis that I offer thee." Margaret responded with a teasing note of her own: "Taking 'P' from 'Page' leaves Age, and you are too old for me." Despite the age difference, they got engaged, and he made a habit of writing poetry for her, inspired by her or about her. Mere days before their wedding, Page wrote another poem about his Peggy, this time while sitting through a boring meeting at the House of Representatives: "The warbling of birds in the grove/The fragrance which flowers send forth/Not so sweet as the voice of my love/And without her can never have worth."

John and Margaret were married and had eight children together. John's seemingly charmed life continued; he became governor from 1802 until 1805 and died in 1808. Eventually, Margaret passed away as well, and the couple was buried together at the family cemetery located on the grounds of the Rosewell Plantation. In 1969, the graves were moved to Abingdon Church, and a stone plaque is the only physical evidence of them to remain on the Rosewell grounds today.

A fire destroyed the house in 1916, leaving only four chimneys; the east wall, with a compasshead window and carved keystone; a few crumbling walls; and the wine cellar intact. The site was professionally excavated

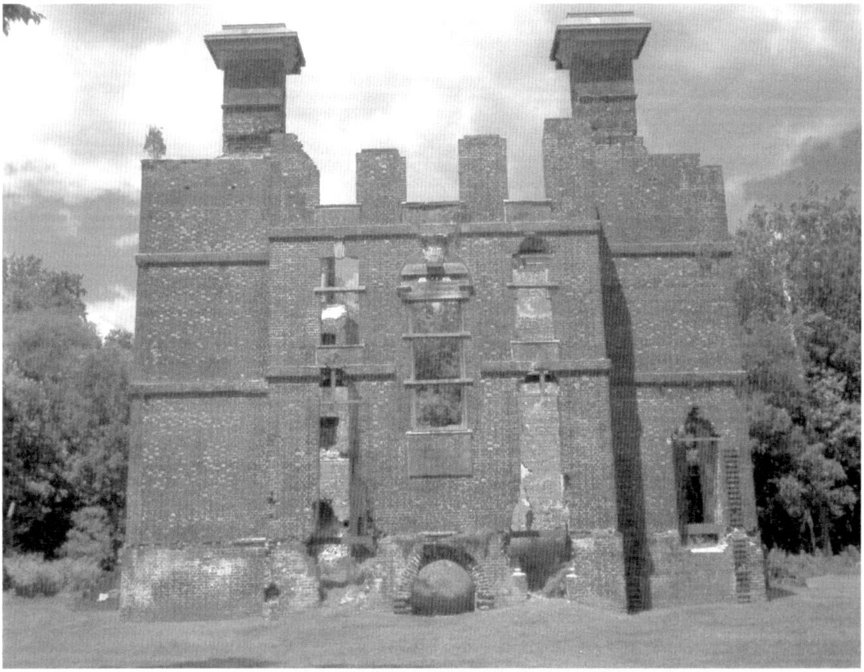

A compasshead window with carved keystone. *Photo by author.*

around the same time as the fire, and the remains of several dead slaves were found buried inside the remaining walls. There is also a fenced-off, thirty-foot-deep ice pit located on the grounds, with old Indian artifacts scattered amongst the rocks and debris.

Only four families have owned Rosewell Plantation over the years, and the last private owners donated the property to the Gloucester Historical Society in 1979. It is currently open at specific times to tour, and a number of tourists have seen and heard ghosts around the ruins, particularly near the front stairs. Mostly, people reported hearing the sounds of slaves returning home from the fields after a long day of work, presumably belonging to the ghosts of the slaves who had been buried inside the walls of the house. Though tourists aren't permitted to enter the ruins, many trespassers have done so. The temptation to explore the ruins is hard to resist because they look like something out of an adventure movie. The wine cellar is particularly alluring, with its dangling vines hanging from the damp ceiling and dark interior. The hollow, underground design of the cellar plays tricks with the acoustics, and the accompanying echoes are strangely muffled. Even without any human voices, the cellar emits a soft mumble that sounds very much like the whisper of a child.

Powhatan's Chimney

We will continue traveling through Gloucester to our next stop, which is another site of strange sounds, Powhatan's Chimney. The chimney is located at Chief Powhatan's former Indian village Werowocomoco, but modern signs incorrectly list it as Wicomico. Powhatan's Chimney is located on Powhatan Drive off Route 17 in a residential neighborhood aptly called Powhatan Chimney. The exact GPS coordinates are 37 17.499, -76 31.877. The chimney itself stands in at least one homeowner's yard close to the road, looking as out of place as Rapunzel's tower.

The story behind Powhatan's Chimney and its place in American history is related to the first American business to fail, the glass factory at Jamestown mentioned in Chapter 3. As was also previously stated, the colonists at Jamestown were chiefly in Virginia to make money for their backers, the Virginia Company of London. The English colonists were instructed to find a water route through America to the Pacific Ocean; to find precious metals like silver, gold and copper; and to produce items for export like spices,

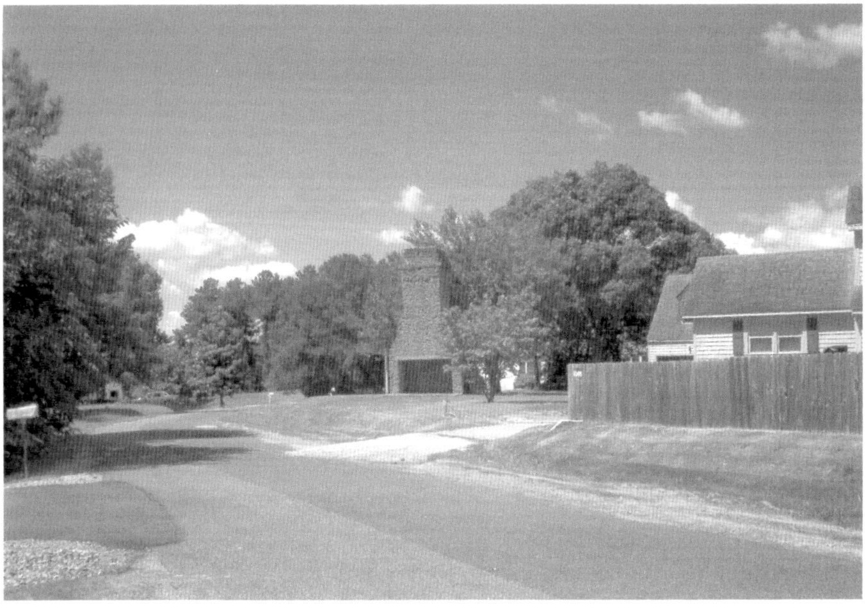

Powhatan's Chimney from the street. *Photo by author.*

lumber, silkworms and glass. Finding a waterway across America proved to be impossible, there were no veins of precious metals to be found and a mulberry tree blight (the only food silkworms eat) during the first year killed all their dreams in one swift blow. The only things left were lumber, glass and spices—until tobacco was discovered—and of those three, only one survived. The lumber and glass trade initially thrived and samples of both were sent back to England, but both businesses soon failed.

If you remember from Chapter 2, Captain Christopher Newport led the English expedition to America before exploring the rivers of southeastern Virginia and returning to England with his findings. Newport then sailed back to America, arriving at the coast of Virginia for the second time in January 1608, later than he had planned. Newport returned to England once again before traveling back to Jamestown as captain of the *Mary Margaret* in September 1608. Also aboard the ship were supplies, a large bed as a gift for Chief Powhatan and approximately one hundred newcomers.

Among the newcomers were a handful of Germans, erroneously described as "Dutchmen," and a Pole. "Dutch" was a corruption of the German word *deutsche*, and the English called all Germans Dutchmen. Two of the Germans were actually Hessians and specialized in glass blowing, and three of the Germans were wainscot sawyers who had been sent to America to produce glass for export and to build houses for the colonists. Four of the eight Germans' names were recorded: Jan Bogdan, the master sawyer Samuel and his two apprentices, Adam and Franz (or Francis). Upon disembarking at Jamestown, the Germans quickly got to work on their assigned tasks; the sawyers built houses for the colonists and prepared planks of lumber to return to England for inspection while the glass blowers set up a furnace and created green-tinted glass to be sent to England as well.

The English settlers had been instructed by the Virginia Company of London to remain on peaceful terms with the natives ("if you neglect this you neglect your safety") because they would need to trade with them. But it seems as if Captain John Smith went against these direct orders and plotted to murder Chief Powhatan. He planned to build the chief a house to go along with his new bed in order to trap him inside it and kill him. Smith volunteered the services of the three German sawyers and the two glass blowers to go to Werowocomoco, Powhatan's main living quarters, to build him a proper English house with a door that locked.

Smith arrived a few weeks later (with George Percy, who recorded the event), dismayed to find the house incomplete, thereby foiling his plans. He returned to Jamestown with the two glass blowers, leaving only Samuel, Adam and Franz to finish Powhatan's house. They had barely completed it, leaving just enough time for Powhatan to try out his key on the front door, when Indian scouts saw Smith, accompanied by a group of colonists, approaching. Chief Powhatan not only abandoned the house, but he also abandoned the entire village, taking the three Germans with him. Smith was furious when he arrived at Werowocomoco and found it deserted. He immediately assumed that the three Germans had betrayed him and defected to Powhatan's camp. From then on, Smith treated Samuel, Adam and Franz as traitors.

Chief Powhatan didn't sit idly by and wait for Smith to kill him; he devised a plan of his own. He sent Franz back to the glass blowing studio near Jamestown Fort in hopes of luring Smith out into the open. The plan worked, and it didn't take long for Smith to arrive to arrest Franz. Unbeknownst to Smith, approximately fifty Indian warriors were hiding in the surrounding woods, waiting to ambush him. An Indian stepped out from behind a tree directly at Smith's back, but Smith turned in the nick of time, and both men fell into the river during the ensuing struggle. Franz readily returned to Jamestown with the drenched Captain Smith. According to Smith's account, he captured Franz and returned him to the fort in shackles.

An Indian leader named Wowinchopunk wasn't so lucky, though. He was captured by the English to be exchanged for the two German prisoners. Wowinchopunk managed to escape, so the English sent a Swiss German mining expert named Volday to retrieve the Germans. Instead, Volday became Powhatan's prisoner too. Volday was eventually released, and he returned to England before coming back to Jamestown, only to be killed by the Indians.

Eventually, Chief Powhatan released Adam, who returned to Jamestown and went unpunished after explaining that he had been a prisoner and not a defector. Adam and Franz were safely inside Jamestown Fort with only their master, Samuel, held as Powhatan's prisoner. Samuel, fearing his days were numbered, escaped but was intercepted by Indians who struck him with their tomahawks until he was dead.

Six months later, during which the Starving Time took place, the colonists abandoned Jamestown and attempted to set sail for England. Adam and Franz chose to remain behind, hoping to retrieve Samuel because they didn't know he had perished. The two Germans traveled to the Indian village to inquire about Samuel. Three days later, on June 10, 1610, Adam and Franz learned that Lord Delaware had arrived at Jamestown, and the colonists, who had never gotten out of the Chesapeake Bay, had returned. Upon hearing the news, Adam and Franz attempted to rejoin the colonists, but the Indians beat out their brains with clubs instead.

Even after death, the three German sawyers have been dishonored by their false historical betrayal. In fact, the only evidence of their hard work and sacrifice is a lone chimney—all that remains of the house they built for Chief Powhatan—that stands in a residential neighborhood of Gloucester. The chimney is all that is left because the wooden house quickly deteriorated after it was abandoned. By the late 1800s, only the chimney remained intact until a storm in 1888 caused even it to collapse. In the 1930s, a new chimney was erected in the same location; it was built with as much of the original material as possible, including bricks and oyster shells. You can see the shells for yourself if you stand on the hearth, which is large enough to accommodate three adults, and look up at the interior chimney walls. And while you are inside the chimney, be sure to listen carefully for any strange sounds.

Residents of the Powhatan Chimney community, in particular those who live closest to the chimney, claimed to have heard odd sounds emanating from it at various times. The sounds vary from a moan to a howl and are most frequently heard during storms, amidst wind-driven rain. One

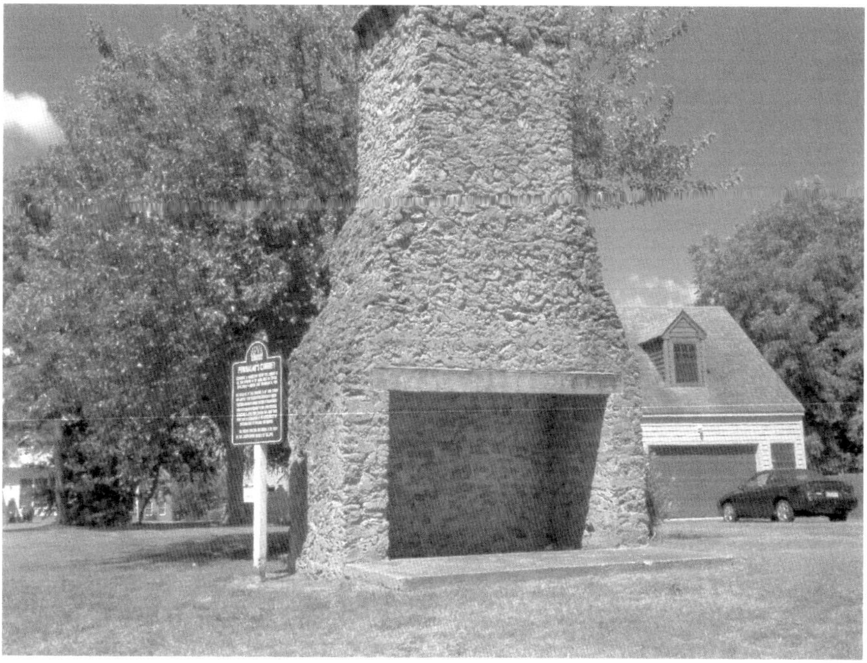

Powhatan's Chimney. *Photo by author.*

obvious explanation for the phenomenon is the wind. At the right speed, coming from the right direction, the wind is forced up through the chimney at a gale-force rate. An alternative explanation, though, is that the sound comes from the spirits of the three German sawyers who bemoan their undeserved fate.

We will leave Gloucester via Route 17 and cross the Coleman Bridge to Yorktown. As you cross the drawbridge, be sure to look across the water to the opposite shore. The beach you see is Yorktown Beach and is close to one of our last stops in the next chapter.

6

YORKTOWN

We will have to cross the York River to get to our next destination, which is the third and final point on the Historic Triangle: Yorktown. Originally, York County was formed in 1632 and was divided into four parishes, with each parish needing its own church. Since then, the area has split into separate cities, of which Yorktown is one. Yorktown saw no fighting during the Civil War but contained the main battlegrounds of the Revolutionary War. Those battlegrounds included but were not limited to the entire downtown portion of colonial Yorktown. Most of the original houses were severely damaged, and even Yorktown Beach saw more than its fair share of fighting. In fact, an event similar to the Boston Tea Party took place off the coast of Yorktown just prior to the start of the Revolutionary War.

Battlefields

There are numerous scenic roads to meander down in Yorktown, like Crawford Road and its bridge, which is best known for the ghosts of three African Americans who were lynched there in the 1800s. Along one such scenic road is our first stop, the Yorktown Battlefields, which were formerly owned by the Yorktown Country Club. The former owner designed the golf course around the existing earthworks, which are still visible today.

Currently, the Battlefields are part of the National Parks System, as is the Yorktown National Cemetery.

The Yorktown National Cemetery was established in 1866 near the spot where the British surrendered in 1781 and Confederate troops marched nearly one hundred years later. An 1868 U.S. Army inspection of the cemetery reported:

> *The interments number 2,180 of which number eleven officers, seven hundred and sixteen white soldiers, four sailors, six colored soldiers, and eight citizens are known and two officers, one thousand four hundred and twenty-two white soldiers, five colored soldiers, and six citizens are unknown. Besides the burial at the cemetery, bodies were removed from Williamsburg in James City County, and altogether from twenty-seven different places in the surrounding country, within the distance of fifty miles.*

Among the eight identified citizens listed, three were Confederate soldiers' wives.

Moore House

There is another cemetery located nearby, much smaller and less known, and we will visit it now. The Moore House is one of the most famous historical attractions in Yorktown, but its accompanying cemetery is what we have come to see. The Moore House is located on Route 238 on the corner of Moore Lane and Hamilton Road. The house is a two-story white building with a tall chimney on each end and the original foundation still intact. The building itself is rather nondescript; one of the small outbuildings is more interesting in appearance because it looks like nothing but a roof protruding from the ground.

The property on which the house now stands was originally established in the 1630s, when it was called York Plantation, but the house itself wasn't built until 1709. The deed changed hands several times before Augustine Moore acquired it in 1767. Moore had inherited three plantations from his father prior to purchasing York Plantation, which was later named for him.

The Moores were upstanding members of the community, with Moore working as an apprentice to William Nelson before becoming a partner of Thomas Nelson Jr. & Co. in 1773. The family lived in the Moore House

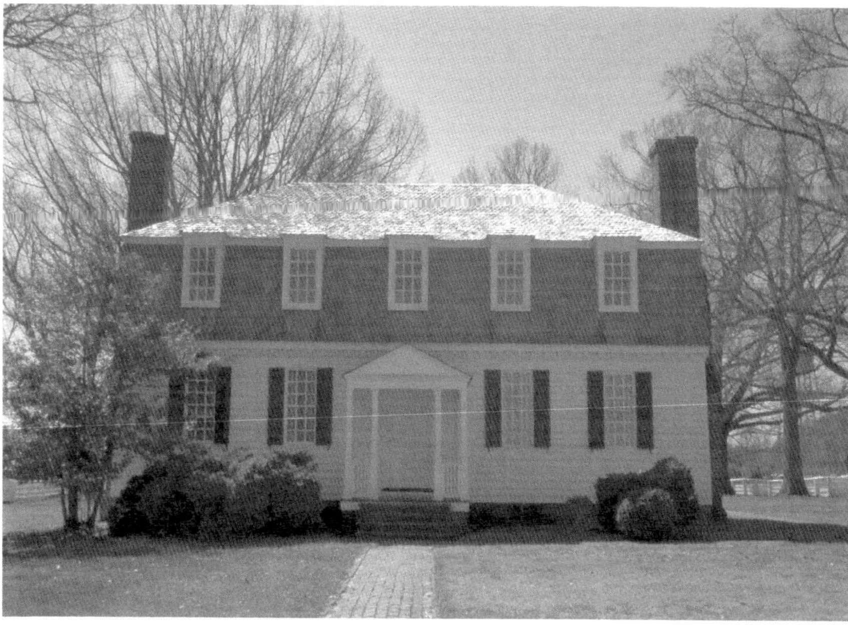

The Moore House. *Photo by author.*

until the start of the Revolutionary War, when a stray bullet killed the eldest son, Augustine Moore Jr., while he tended the family fields. He was buried at the family cemetery on the property, and the surviving Moores fled to Richmond after the commencement of his funeral service.

The Revolutionary War continued to grind on around the Moore House, which had been found abandoned and was then used as a hospital. Soon another fatality occurred near it. John Turner, a curious onlooker, was accidentally wounded by a musket ball near the Yorktown Battlefields as he watched the skirmish from what he thought was a safe distance. Turner's wife, Clara, was nursing wounded soldiers at the Moore House when her husband staggered up to her, mortally wounded. He died in her arms and was buried at the Moore family cemetery. His devoted wife wrote a poem for him and had it inscribed on his tombstone:

A cruel ball so sudden to disarm
And take my tender husband from my arms
How can I grieve too much what time shall end
By mourning for so good, so kind a friend.

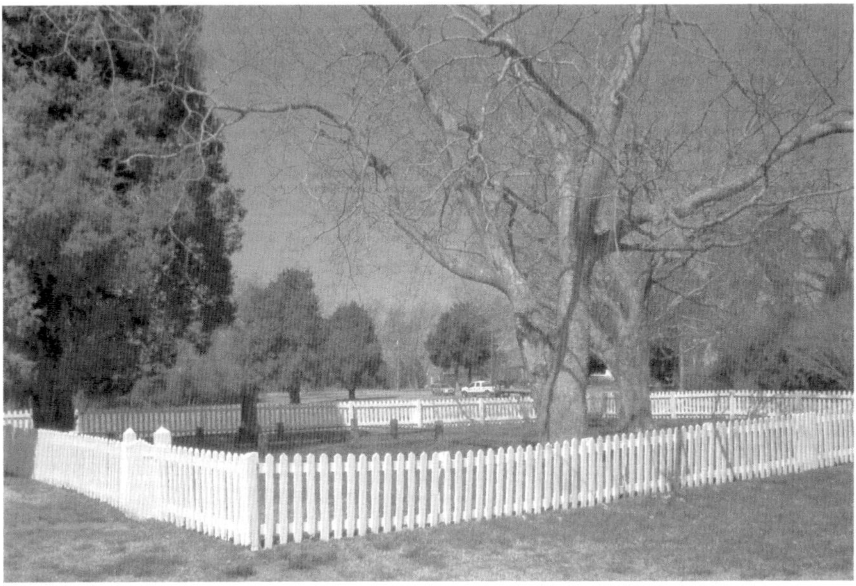

The cemetery at the Moore House. *Photo by author.*

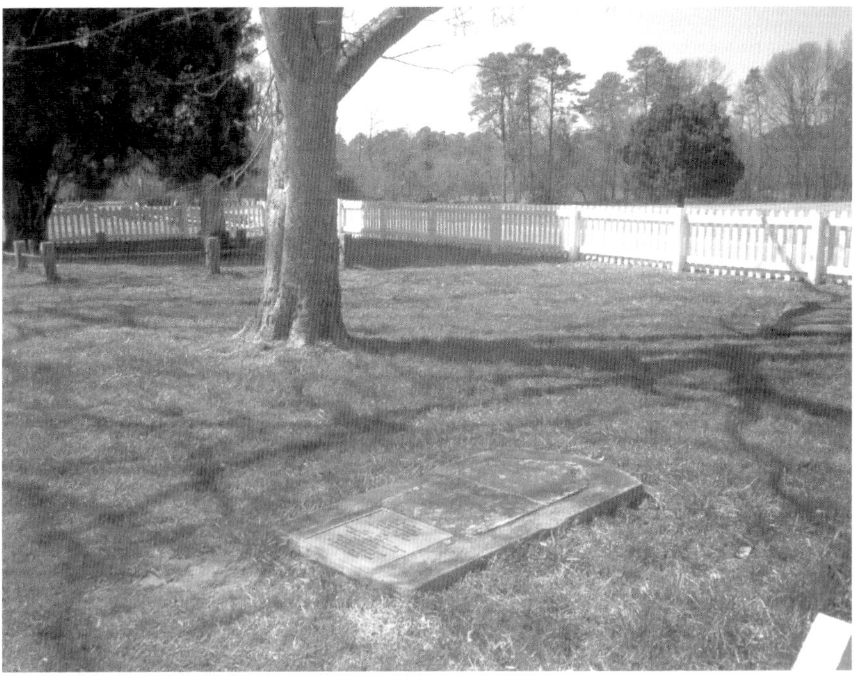

A Moore House tombstone. *Photo by author.*

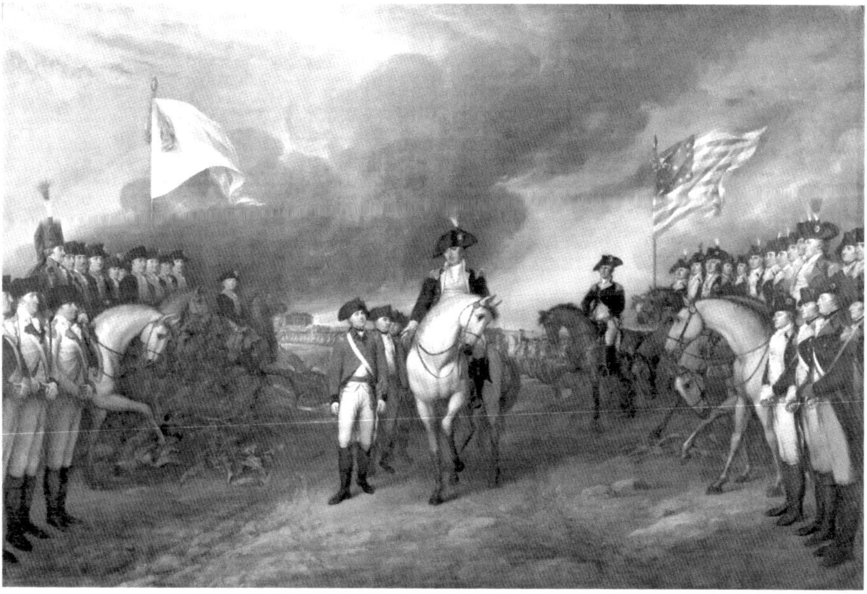

Lord Cornwallis surrenders. *Photo courtesy of the Library of Congress.*

The Revolutionary War ended when Lord Cornwallis sent a note to General George Washington that read, "Sir, I propose a cessation of hostilities for twenty-four hours, and that two officers may be appointed by each side, to meet at Mr. Moore's house, to settle terms for the surrender." The Moore House was chosen because it had been out of the line of fire and was therefore relatively undamaged. It was also considered to be a neutral location and convenient for both sides. Negotiations of surrender did indeed take place the following day, on October 19, 1781, inside the Moore House in a room that is now referred to as "Surrender Room."

The house remained in the Moore family until 1797 and then changed hands many times until the National Parks Service acquired it in the 1930s. It restored the house, and the original seventeenth-century foundations of three other outbuildings were subsequently discovered. One of these outbuildings, an icehouse, is the aforementioned roof protruding from the grassy yard. Over the years, strange voices have been heard and recorded coming from inside the house, and a ghost was seen napping in an upstairs bedroom. Part of a 1984 TV series called *George Washington* was filmed in the Moore House, and several crew members heard the sound of water running and toilets flushing, although there isn't any plumbing inside the house.

General Nelson's House

Our next place to visit is another haunted house, that of General Nelson, located on the corner of Main and Nelson Streets in downtown colonial Yorktown. Much of the three-story brick building is original, dating to 1711 or 1715. There are several holes visible in the outer walls of the house facing Nelson Street; the smaller ones were made by musket balls, and the larger ones were made by cannonballs during the Revolutionary War. However, the cannonballs inserted in the holes are not the originals; the last private owner inserted real Civil War–era cannonballs into the indentations made during the Revolutionary War.

The house also contains a disused secret stairway leading from the first-floor hallway to a third-floor garret and a hidden cubby in the dining room. These concealed features are the heart of the following stories and one of the main reasons we are visiting the house today. Another reason we are here is to see its historical contents; the house contains everything from simple colonial children's toys to repaired "ventilation holes" in the doors, used to

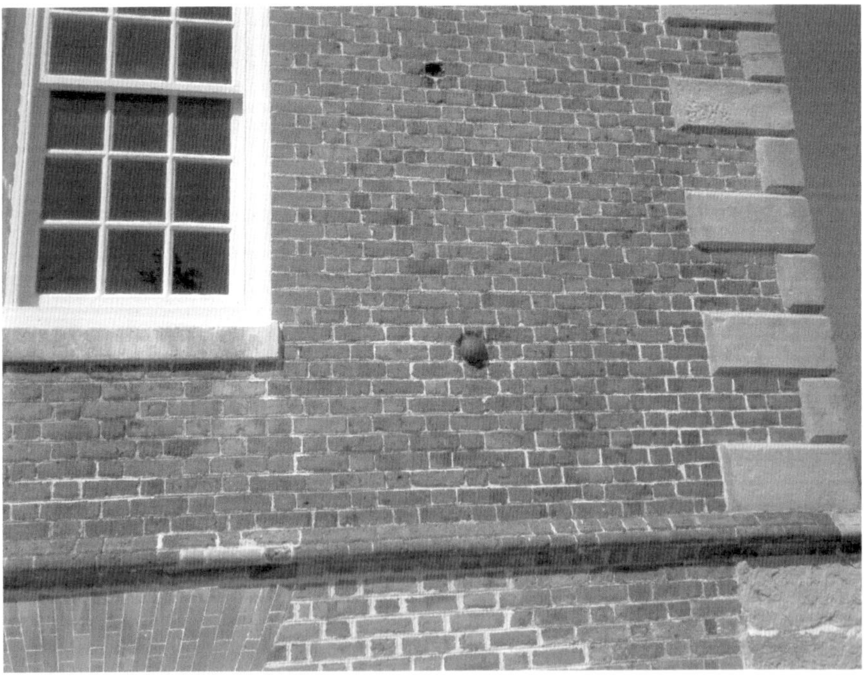

Cannonballs in Nelson's walls. *Photo by author.*

air out the stench of rotting flesh during the Civil War. Yet another reason to visit the house is to learn about its former owner and namesake, General Nelson. Thomas Nelson Jr. was the secretary for the Virginia Colony, a signer of the Declaration of Independence, a major financial contributor to the Patriot cause and the commander of the Virginia militia troops during the 1781 siege of Yorktown. He was also an avid gardener in his free time; his three-foot-long cucumbers were even mentioned in a 1738 edition of the *Virginia Gazette*.

General Nelson lived in the house from 1738 until 1789 but was forced to abandon it at the beginning of the Revolutionary War. Unlike the Moore House, the enemy British soldiers confiscated the empty Nelson House, and Lord Cornwallis himself used it as his headquarters during his siege of Yorktown. In late September 1781, George Washington, along with Thomas Nelson Jr. and an army of Patriot troops, was preparing to bomb Yorktown. Washington knew that Nelson was from Yorktown, so he asked him where they should start bombing. Nelson, without hesitation, pointed out his own house, where Cornwallis was staying, and said it would be a fine place to start. Washington obliged, and serious shelling ensued, hitting the secret passageway in the Nelson House and killing a British soldier who had been hiding within it. Lord Cornwallis was unharmed, perhaps because he'd been hiding in a nearby cave, which we will visit momentarily.

After the Revolutionary War was over, the Nelsons returned to their house and rebuilt it, but Thomas Nelson Jr. had little time to enjoy it. He died during an asthma attack six years later. Incidentally, he was never reimbursed for the contributions he had made during the war effort. He was buried, along with several generations of Nelsons, at the nearby Grace Episcopal Church, which we'll visit next. The house remained in the Nelson family until 1908.

But before the house left their hands, the Nelsons were forced to abandon it once again, this time due to the Civil War. Though no actual fighting took place in Yorktown, most of the houses were abandoned and used as hospitals, suffering tremendous damage as a result. The Nelson House was no different, and circular holes were cut into various doors to aid in ventilation, with the patches still clearly visible to this day. The most critically wounded patients were housed on the third floor of the house, with the majority of them dying after their amputated limbs became gangrenous. These same amputated limbs were piled up around the house, and the smell of rotting flesh was so foul that all of the windows were constantly left wide open,

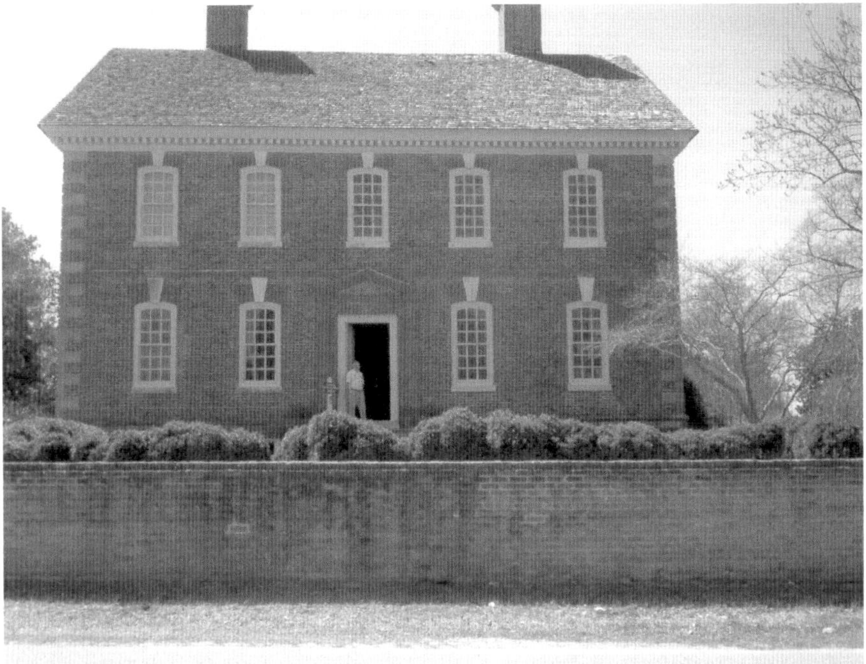

Thomas Nelson Jr. at Nelson House. *Photo by author.*

regardless of the weather. Even with the windows left open and ventilation holes cut in the doors, the putrid smell was overwhelming at best.

The house left the Nelson family in 1908, and the last private owners were the Blow family. One day, Mrs. Blow was hosting a luncheon party when suddenly a secret panel in the dining room flew open and something invisible crashed into the sideboard. All the guests witnessed this strange occurrence, and none of them believed Mrs. Blow's explanation that it had merely been a draft. The National Parks Service acquired the house from the Blows in 1968, and since then, a number of strange events have been reported. Many tourists have seen the face of a man peering out of a third-floor window when no one was in the house. Former employees have seen the dark figure of a man in the vacant third-floor halls. The spirit is believed to belong to the unfortunate British soldier who was killed in the secret stairwell during the Revolutionary War.

Perhaps you will get lucky and Thomas Nelson Jr., a direct descendant of the first Thomas Nelson Jr., will be your tour guide. He will be more than happy to show you the hidden cubby in the dining room, the very same one that blew open during Mrs. Blow's luncheon. It is accessed through the china sideboard, where shelves can be rearranged to reveal a small doorway that leads to a space large enough to conceal four people. Likewise, he will show you where the secret stairwell is located; he will even open the door and give you an invitation to poke your head inside.

Grace Episcopal Church

We will continue our tour on foot and visit the grave of Thomas Nelson Jr. at the neighboring Grace Episcopal Church, located at 111 Church Street. The church is a small, buttercream-colored building that was originally made from irregularly shaped marlstones, a combination of shell, lime and clay found along the York River. The church has few windows, completely lacking in stained glass or other ornamentation, and a squat steeple atop the roof.

The first church was built in 1642 at the present-day Coast Guard Reserve Training Center, where Major William Gooch was buried in 1655. His tombstone reveals how much the church meant to him:

Within this tomb there doth interred lie, No shape but substance true nobility;
Itself though young in yeares but twenty-nine,
Yet graced with vertues morall and divine;
The church from him did good participate;
In councill rare fit to adorn a state.

The second church was built in 1667 atop the foundations of the first church. The third church, and the one you are visiting, was built in 1697 after Lieutenant Governor Frances Nicholson pledged "twenty pounds sterl" if the church was built within two years. William Byrd, from the Westover Plantation section in Chapter 5, attended this church on occasion, taking a boat down the York River to arrive in time for services. Three generations of Nelsons were active members of the church and are buried at its cemetery, including Thomas Nelson Jr.

The British army decided to use the church, which had a separate bell tower by then, as a gunpowder storage facility when it invaded Yorktown during the Revolutionary War. The church managed to survive only to be burned down later, on March 4, 1814. Mrs. Gibbons's house caught

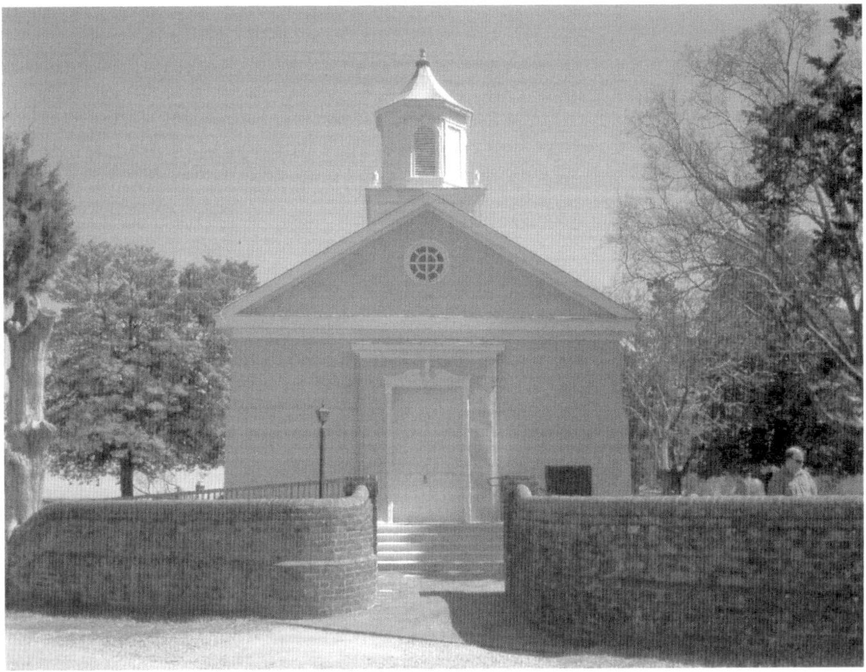

Grace Episcopal Church. *Photo by author.*

fire and spread "like a train of powder," burning down the courthouse, Nelson House, the church and the whole town below the hill that was occupied by "poor people, who are now thrown out upon the world without a shelter or a cent to aid them." In 1848, the church was rebuilt due to the perseverance of the German immigrant Reverend Dr. Charles Minnigerode, who had also been a Latin and Greek professor at the College of William and Mary. Reverend Minnigerode is also credited with introducing to his friends the custom of decorating an indoor evergreen tree in the house at Christmastime.

After being rebuilt, the church remained intact until the Civil War. Confederate soldiers used the church at the beginning of the war, and the soldiers who were quartered there used the long, flat gravestones in the cemetery as their mess tables. But the Confederates were forced to retreat from the Union army, which soon arrived in Yorktown. The Federal troops decided to store their gunpowder in the nearby courthouse instead of a distant cave like the Confederates had done. As a result, the courthouse inadvertently blew up, accidentally damaging the church in the process. The church bell, originally cast in 1724–25, fell and cracked and was later carried north by the Union army. It was discovered at a smelting company in Philadelphia, where it was recast and sent home to the church in 1882.

At the start of the Civil War, a fourteen-year-old resident of Yorktown, Richard Benjamin Gilliam, managed to enlist in the Confederate army. His age was soon found out, though, and he was sent back home to his relieved mother. As soon as Gilliam turned seventeen, he promptly reenlisted and was stationed at a base in nearby Gloucester. By then, Gilliam had a girlfriend who lived in Yorktown, and only the York River separated them. One evening, Gilliam decided to ride his horse across the vast river (the ferries used to transport people, horses and wagons didn't cross the river at night) to pay her a surprise visit. Gilliam barely made it to the other shore when, exhausted, he stumbled on the severely damaged Grace Episcopal Church. He dragged himself inside the desolate church, his girlfriend completely forgotten. Instead, he laid his weary body down on a pew and fell asleep. Bright and early the next morning, parishioners arrived for services and began to pray aloud, asking the good Lord to help with the war. Gilliam, still sprawled unseen on the pew, called out, "Who's there and what do you want?" The frightened parishioners fled from what they thought was the voice of God, and tales of the church itself being haunted first began to circulate.

The church fell into disuse until the early 1870s, when Yorktown residents began to return to their abandoned homes. The church was quickly rebuilt, a reverend was installed and members returned to the fold. A former professor at the College of William and Mary, Reverend Dr. L.B. Wharton, began preaching at the church in 1875. He continued working at, on and around the church for the next twenty years without pay. He died in 1907 at the Williamsburg train station while buying a ticket bound for Richmond to attend a Confederate reunion. His obituary stated, "His reunion was with the greater hosts of the dead."

The church has remained active with a flourishing membership ever since 1887. Some restoration to the church took place in the 1920s; electricity was installed in 1927, though the church had no fixtures until Reverend C.A. Pruden insisted on it. To help raise money, he began giving tours to visitors, raising the annual donations from less than $400 to over $1,000, more than enough to cover the first electric bill in March 1934: a grand total of $1.50. An addition, the parish hall, was made to the church in 1951. During the 1970s, more remodeling took place,

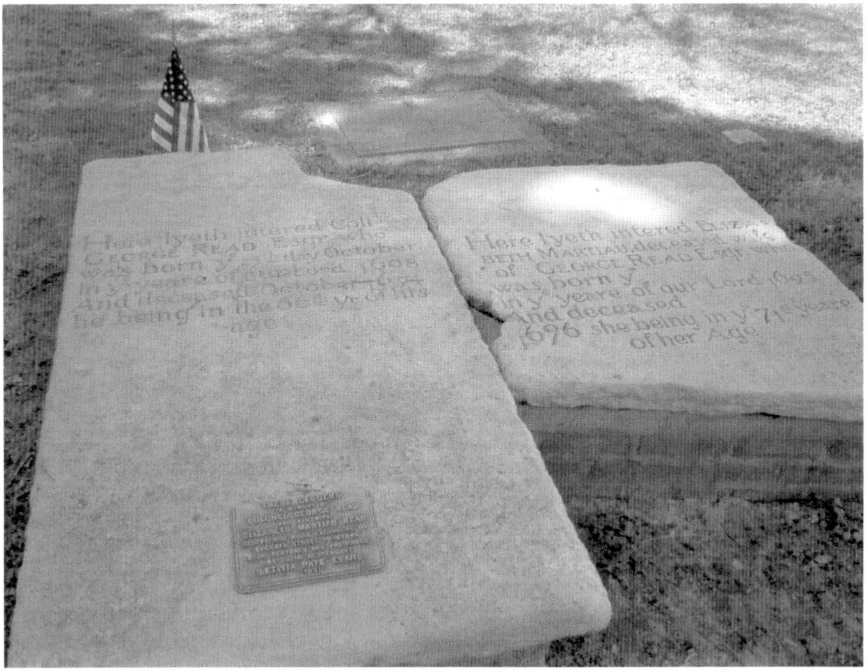

The cemetery at the Grace Episcopal Church. *Photo by author.*

and the original marl walls were skillfully exposed, preserved and then stuccoed. The church is beautifully constructed in its deceptive simplicity, accentuated by its charming color and surrounding cemetery.

The Grace Episcopal Church is one of the oldest churches in America, but there is more here than meets the eye; the thriving church has more members now than ever, and it has received numerous donations of religious folk art over the years. It even offers a canvas labyrinth tour on special occasions, Halloween in particular. An Easter parade that travels down Main Street, with participants carrying folk art crosses, has become a tradition. The church is home to a set of 1649 communion silver—a flagon and a chalice—forged in London. At one time, the church possessed a silver paten after Martin's Hundred merged with York in 1712, but now, the silver paten is kept at St. John's Episcopal Church in Hampton, which we will visit in Chapter 7.

CORNWALLIS'S CAVE

There's yet another historical site in Yorktown associated with the Revolutionary War, and we will visit it next. It faces one of the cleanest and best-tended public beaches on the peninsula, Yorktown Beach. You can walk or drive down the steep hill toward the beach to Water Street, where you will come to Cornwallis's Cave, one of the most famous caves in America. Lord Cornwallis himself used the cave as a lookout point and a hideout during his siege of Yorktown, and he constantly kept a guard posted there. Perhaps Cornwallis even hid in the cave when his headquarters at the Nelson House was being bombed.

You can no longer explore the inside of the cave due to its barred entrance, erected because of the damage being inflicted on it by littering and vandalizing locals and tourists. The small, natural sandstone cave dates back to at least the early 1700s, when it was first recorded because pirates, possibly even Blackbeard himself, used it as a hideout. But it is believed the cave dates back even further, with Native Americans using it throughout

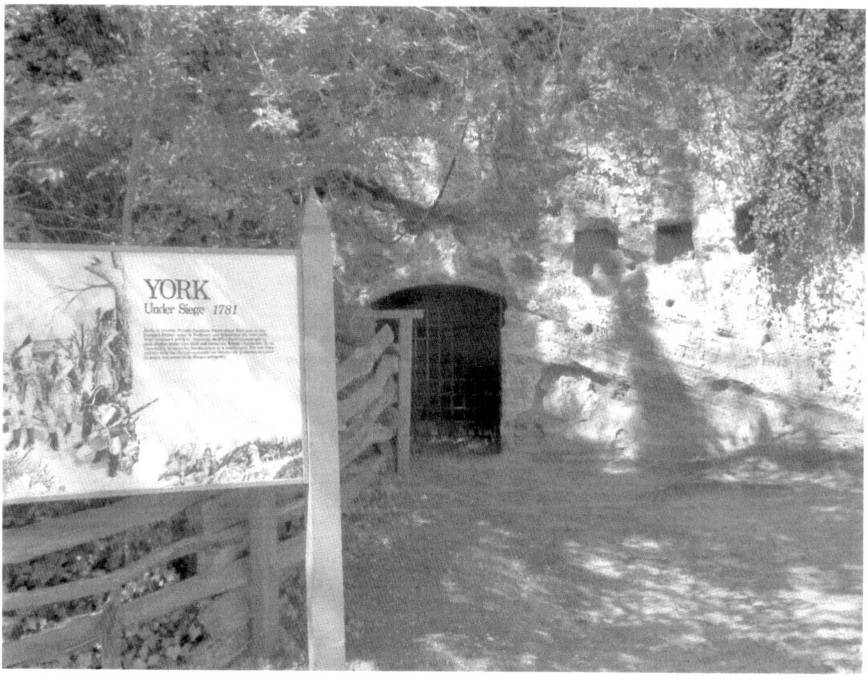

Cornwallis's Cave. *Photo by author.*

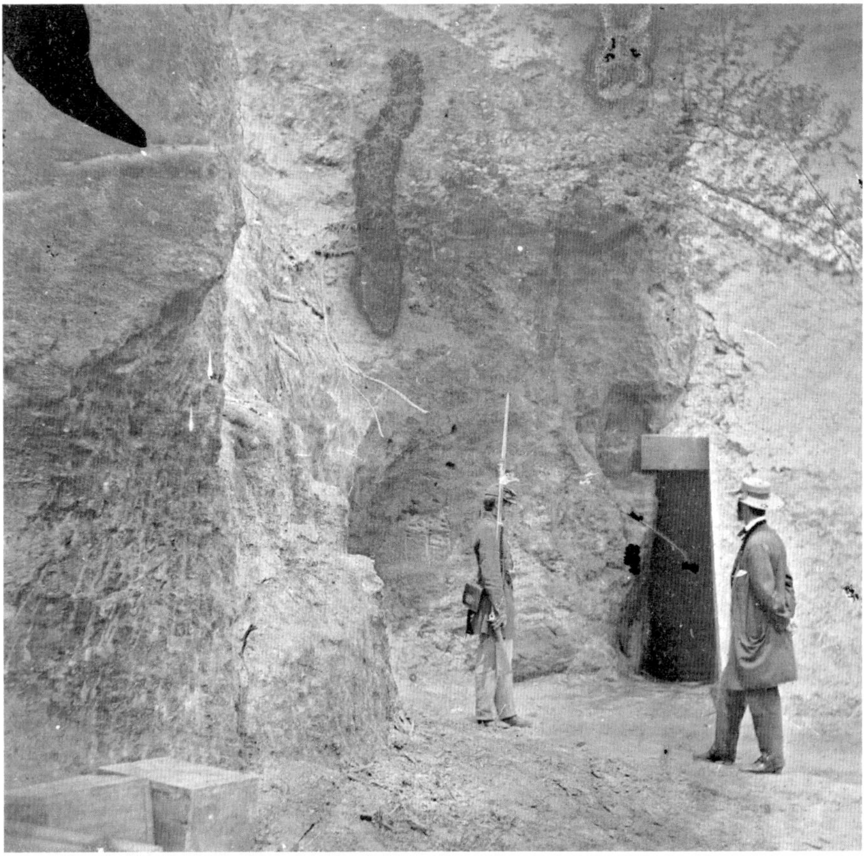

Cornwallis's Cave during the Civil War. *Photo courtesy of the Library of Congress.*

the 1600s, predating both the English and the pirates. It is possible that Pocahontas and her childhood friends played in the cave, shouting into it just to hear the echo. The cave was well known by the 1800s, and as previously mentioned, the Confederate army stored gunpowder in it during the Civil War.

Near the end of the Revolutionary War, in mid-October 1781, many of Cornwallis's men hid in the cave because they knew they were surrounded and losing the battle. Some of the British soldiers hiding in the cave decided their only chance of survival was to board a few small boats and paddle their way across the York River to safety. No sooner did the unsuspecting soldiers get out into deep water than a sudden squall arose with a blinding rain and driving winds, churning the river into towering walls of water. Most of the soldiers at sea were killed by the pounding surf, and some of the men cowering inside the cave were drowned in the high tide.

Since then, especially when an unexpected squall arises on the York River, people have heard the sounds of screaming coming from inside the barred cave. Tourists and neighbors have been out for a stroll along the beach only to be caught in a storm unawares, and they sought refuge inside the cave before it was barred. Once inside, they were greeted by unseen voices calling for help along with moaning and shrieking sounds. A couple of local residents actually recorded some very strange sounds coming from inside the cave. As with Powhatan's Chimney in Chapter 5, perhaps it is just the sound of the wind howling through the cave…or perhaps it is something a bit more supernatural.

We will leave Yorktown and its Revolutionary- and Civil War–era curiosities and travel southeast to the city of Hampton. Humans have inhabited Hampton since at least the mid- to late 1500s, and it has been

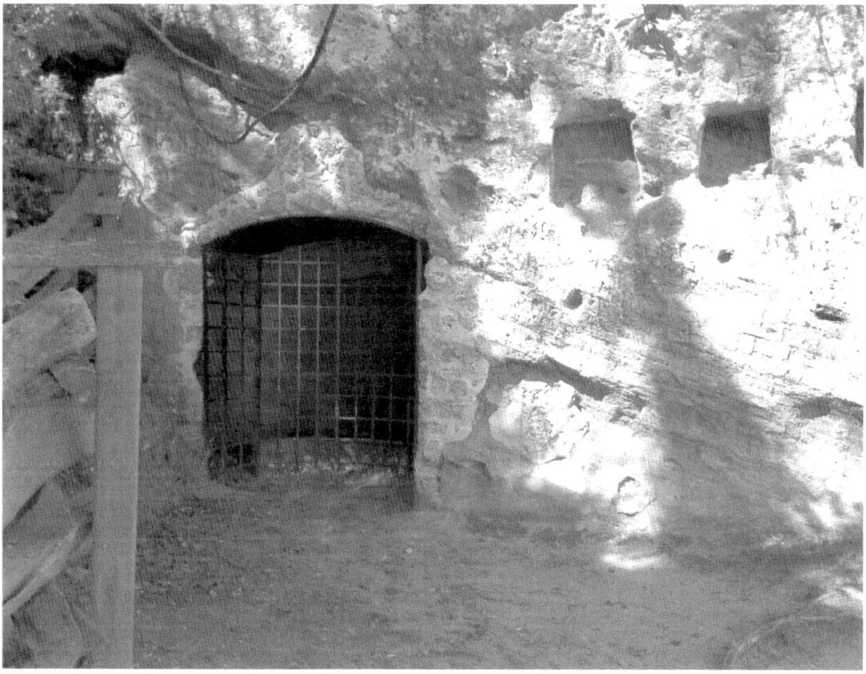

The barred entrance to Cornwallis's Cave. *Photo by author.*

inhabited by an English-speaking population since the early 1600s, making it the oldest surviving English city in America. In Hampton, you will learn more about the Algonquian Indians, visit the cemetery where the first African American born in America is buried and see the first American military base.

7

HAMPTON

Hampton was called Kecoughtan by the Algonquian Indians who first lived there, and the city is the oldest English municipality in America. The English settlers who arrived in the Hampton Roads area in 1607 decided that Kecoughtan should be included in their newly acquired James River District. The district also included Jamestown, Williamsburg and Yorktown—the entire Historic Triangle as we know it today. This area, along with the rest of the James River District, made up practically the entire southeastern peninsula of Virginia, which is now part of Hampton Roads.

When the English first settled in Kecoughtan, they renamed it Elizabeth City, which was formally established in 1610. The Native American inhabitants were gradually pushed off their own land and completely relinquished their hold over the next forty years. Elizabeth City eventually became Hampton, and the boundaries of the city have changed over the years, as has its name. Regardless of the name, though, Hampton was the first permanently established city in America, since even Jamestown was temporarily abandoned.

The War of 1812 formally began at the port of Hampton, close to where Blackbeard's head was posted, when the British army invaded on June 19, 1813. For two days straight, over two thousand British troops attempted to come ashore but were repeatedly thwarted and finally forced to retreat altogether. They returned in full force three days later; this time, they managed to come ashore and make quick work of destroying the entire city of Hampton. The British soldiers burned practically every building they came across, homes, offices and businesses alike.

They also raped most of the Hampton women they encountered. Major Crutchfield was an eyewitness to the carnage, and he reported to Governor Barbour that "the unfortunate females of Hampton who could not leave town were suffered to be abused in the most shameful manner." It seems no woman was safe, from teenaged daughters to geriatric grandmothers of all social classes.

The ruthless British troops even killed a man who had lain in his deathbed for six months prior to their invasion, and then they shot his wife "in the hip at the same time, and kill[ed] his faithful dog lying at his feet." The next year, the British army invaded the White House, causing President James Madison and First Lady Dolley Madison to flee for their lives. Another year later, Andrew Jackson won the final battle of the war in New Orleans, and the war was officially over in 1815.

Hampton barely had time to recover and rebuild, bigger and better than ever, before it was set on fire again in 1861. This time, the Confederate army burned the city in a vain attempt to prevent the Union army from looting it. Apparently, their motto was "If we can't have it, neither can you," and they did a good job of enforcing it. The city was rebuilt again after the Civil War was over, and it became even more bustling and booming than ever. Today, Hampton looks similar to how it did at the turn of the twentieth century, though it is much, much larger, with a vast population of military personnel and shipyard workers. It is still as nice a place to live as it was when Captain John Smith first visited. "We were never so merry…than in the dry, warm, smoky houses of Kecoughtan," he wrote in 1608.

Langley Air Force Base

Our first stop in Hampton is part of Langley Air Force Base, or LAFB for short. It was established in 1916 and merged with the army base Fort Eustis in 2010. LAFB is located next to the NASA Research Center, and the only way to gain admittance is with a military ID. However, we will be visiting one of its housing facilities called Bethel Manor. Bethel Manor is technically part of York County, although it belongs to LAFB in Hampton. It is located just off Hampton Highway (Route 134) and is where the Civil War Battle of Big Bethel took place, one of the strangest battles to have occurred during the war.

After Fort Sumter surrendered to the Confederates on April 14, 1861, the state of Virginia seceded from the Union. Within days, the Union army took control of Fort Monroe in Hampton, which we will be visiting soon. Confederate colonel John B. Magruder quickly set up headquarters in Yorktown, and in the first week of June, he sent Confederate troops to Big Bethel Church and Little Bethel Church to harass the Union army at nearby Fort Monroe.

In retaliation, the Union army attacked the Confederates at dawn on June 10, 1861, hoping they would retreat. Instead, fifty-eight Union soldiers were wounded and eighteen were killed, mostly by fellow Union soldiers. Two Union regiments mistook each other for Confederates and opened fire on one another, thereby alerting the real Confederates to their proximity. So the Confederates were ready when another Union regiment attempted to flank them. A different Union regiment withdrew too early, forcing the remaining one to withdraw as well. Sensing a dismal failure, the Union army gave up and returned to its bases. All in all, only seven Confederate soldiers were wounded and only one was killed that day.

The location of the actual battle is now submerged under water or has been developed. All that commemorates this epic battle is an engraved

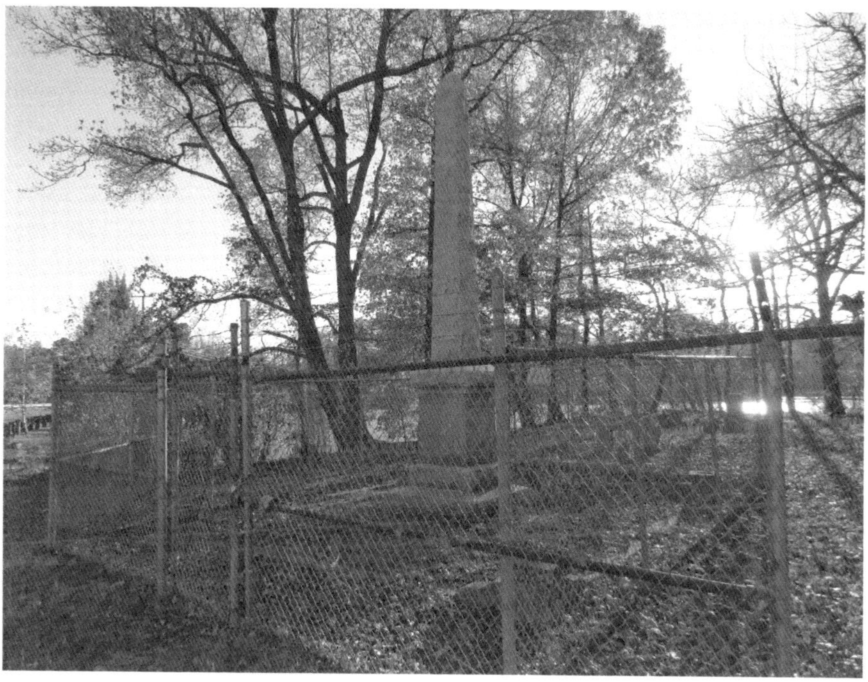

The Bethel obelisk. *Photo by author.*

obelisk at Bethel Cemetery, erected in 1905 by the United Daughters of the Confederacy. It reads: "To commemorate the Battle of Bethel, June 10th 1861. The first conflict between the Confederate and Federal land forces and in memory of Henry L Wyatt, Private, Co A 1st Regiment North Carolina volunteers. The first Confederate soldier to fall in actual battle."

The next three places we will visit in Hampton are all cemeteries, among the oldest in America. Therefore, they are all three steeped in history and legends. Our next stop involves Captain William Tucker, who owned the first African American baby born in America.

TUCKER FAMILY CEMETERY

Captain William Tucker is perhaps most famous for ending peace talks with the local Indians after the Massacre of 1622 by proposing a toast laced with poison prepared by Dr. John Potts. Chief Opechancanough, brother of the late Chief Powhatan, accepted the drink, took a few sips and passed it on to prominent members of neighboring tribes. Not all the Indians died immediately, Chief Opechancanough being one of them, but the survivors were shot by Englishmen hiding nearby. All in all, over two hundred Indians died that day, and it is believed that their remains were buried in a mass grave along present-day Shore Drive.

Despite being poisoned and shot, Chief Opechancanough escaped, living to fight the English another day. And fight he did, staging the Massacre of 1644 at the age of ninety, which resulted in Captain Tucker's murder and Opechancanough's capture in 1646. The chief was paraded around Jamestown before he was shot in the back by the soldier who was guarding him. After the Massacre of 1644, it was considered a felony for an Englishman to go to Gloucester because of the Indian village Werowocomoco and Opechancanough's tribe near West Point along the Pamunkey River.

Yet Captain Tucker is less known for another historical fact—owning the first African American baby born in America. The baby was named after his owner William, born to Anthony and Isabella, last names unknown. Anthony and Isabella came to America as "indentured servants" but arrived in this country on a Dutch slave ship in 1619. It is believed that Anthony and Isabella were among the first twenty Africans brought to America by Captain Jope aboard the *White Lyon*, as recorded by John Rolfe. Although they were supposed to be indentured servants, Anthony and Isabella were sold to Captain Tucker from the auction block at present-day Fort Monroe, and they remained in his service until their deaths.

Not only was Captain Tucker the commander of Kecoughtan, but he was also an accomplished sea captain, adept at navigating a ship for long distances. He was also an extremely wealthy plantation owner in Virginia. Captain Tucker; his wife, Mary Elizabeth Thompson; their family; and their newly acquired slaves settled at Jamestown for a while before moving east to the property where Bluebird Gap Farm now stands, which we will be visiting next. After the 1622 Indian Massacre, Anthony and Isabella were recorded among the living.

It is uncertain whether Anthony and Isabella were a couple, legitimately married or otherwise, before or after coming to America. However, it is documented that Isabella gave birth to the first African American baby born in America on January 3, 1624. Captain Tucker had the baby baptized and bestowed his last name on the child, which would become a common slave tradition in later years. In the official 1624 census, the Tucker slaves were listed as part of Captain Tucker's personal property, along with forty barrels of corn, four pistols and three swine.

In 1626, Captain Tucker was still the commander of Kecoughtan, and he moved to an area near the present-day Willoughby Spit. He sold this property in 1628, bought more property in Elizabeth City in 1632 and lived there for a while. He continued buying property and land patents in southeastern Virginia throughout the 1630s and then traveled to Ireland, England and back to America before his death during the Indian Massacre of 1644. In his will, Captain Tucker left all of his American property to his biological son, who was also named William.

William Tucker the slave was eventually freed and went on to marry a biracial woman. Thomas Tucker was the last slave in the Tucker family to be freed, and he vowed to never work for a white man or woman ever again. And he never did, though he started out his first business, a moving company, with little more than a dream and an ox-drawn cart. The Tucker descendants include the Williamses; Thelma Williams was one of the last Tuckers living in Hampton until her death in 2006.

Captain William Tucker is believed to have been buried at a private cemetery located on the grounds of some property he once owned in Hampton, near present-day Bluebird Gap Farm and Hampton University. Anthony, Isabella and their son were all buried at this cemetery, but many of the seventeenth-century graves are unmarked. In 1896, the African American Tucker's descendants bought the deed to the cemetery for $100. It

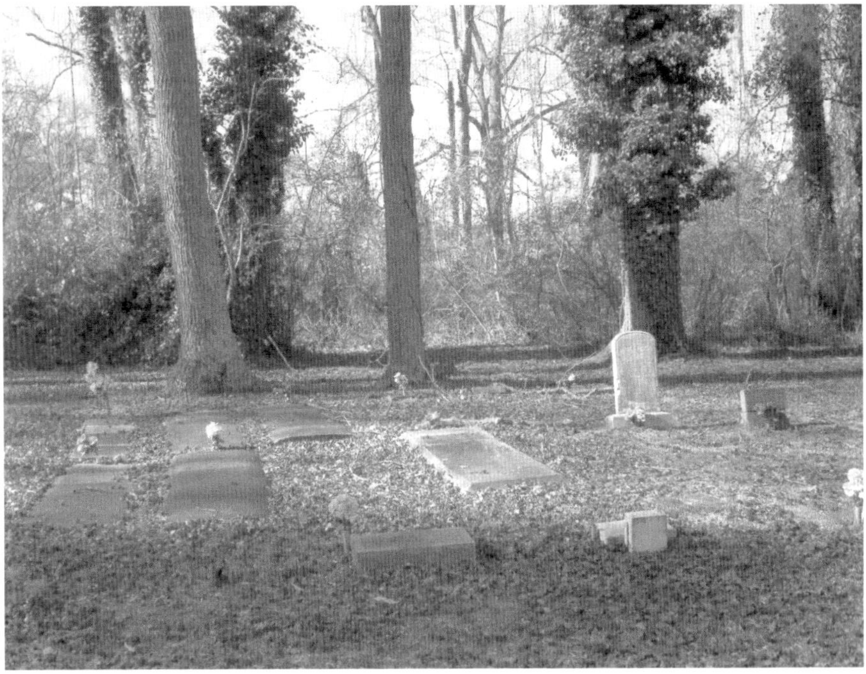

Above: Tucker Family Cemetery. *Photo by author*.

Below: The Tucker Cemetery entrance. *Photo by author.*

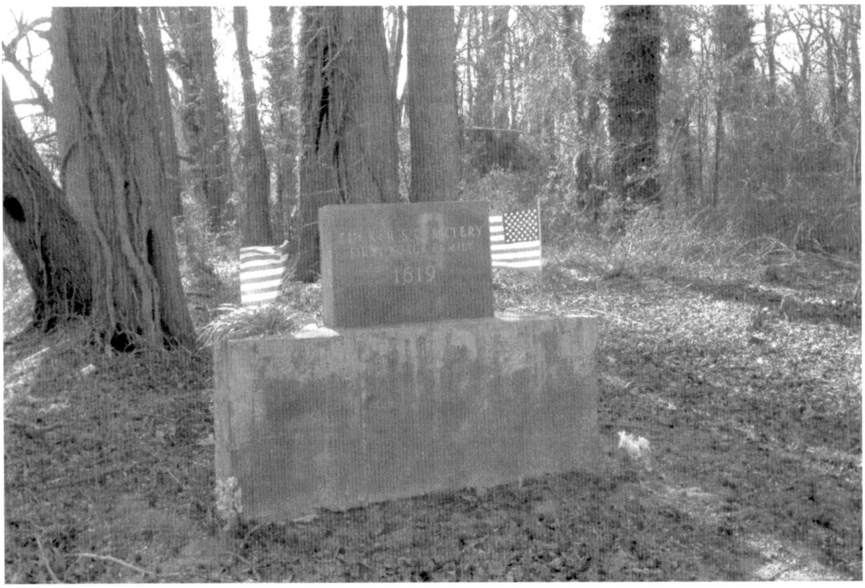

was referred to as "the old colored burial ground" on the document. Thelma Williams was even interred there after her passing, and her headstone is easily identifiable. To date, a total of twenty-four interments have been recorded, with many more undocumented, like Captain William Tucker and his African American namesake.

Currently, the Tucker Family Cemetery is the oldest African American cemetery in America and is located at the corner of Cornelius Drive and Sharon Court near Power Plant Parkway, smack-dab in the middle of a modest residential neighborhood. In fact, the cemetery practically runs right through several different backyards with a thick copse of trees on the other side of it. The cemetery is easy to miss; you could drive past it dozens of times and have no idea that a nearly four-hundred-year-old cemetery lies within a stone's throw. A simple granite marker at the entrance peeks out from between the houses and provides the only visible warning from the street that you have reached your destination. You should park along the street and walk right on in after you find it. Lacking a fence, it is quite easy to access.

Bluebird Gap Farm

We will leave this ancient graveyard to its peaceful slumber and head to the next cemetery on our tour of Hampton Roads, the Davis Family Cemetery at Bluebird Gap Farm. This stop also involves Captain William Tucker, who was much more than the slave-owning commander of Kecoughtan. Children will especially enjoy this site, as it is located at a small, operational farm equipped with a playground. The Davis Family Cemetery can be found off Azalea Trail on the grounds of Bluebird Gap Farm, located at 60 Pine Chapel Road next to the Hampton Coliseum, just off I-64.

The Davis Cemetery is small, measuring approximately twenty-five feet square, and is enclosed by a rusted fence. The cemetery is surrounded by woods and overlooks the marshiest portion of Newmarket Creek. It is located well off the beaten path on the grounds of the farm and contains fewer than a dozen graves. The marked graves date back to the early 1800s, before the cemetery was formally established in 1908.

The property on which Bluebird Gap Farm now stands has an interesting history dating back to the early 1600s. Captain William Tucker, whom you just met, was the first known owner of the property in 1622. As commander

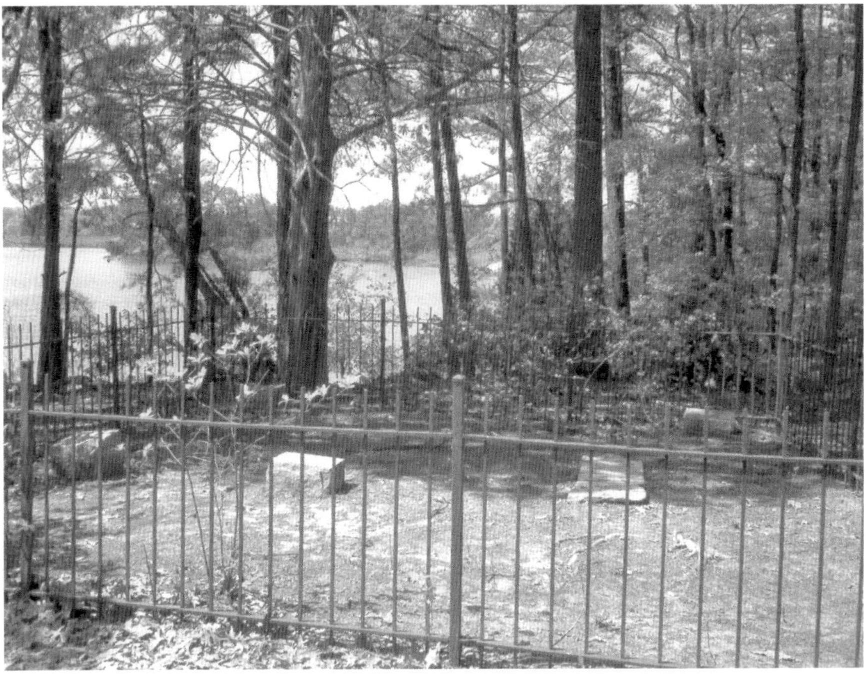

The Davis Cemetery at Bluebird Gap Farm. *Photo by author.*

of Kecoughtan, his post was located a little farther southeast on the peninsula. Captain Tucker moved shortly thereafter to Indian (House) Thicket, located on the grounds of present-day Hampton University, in order to be closer to his post.

In 1645, almost a year after Captain Tucker's death, Ralph Barlowe bought part of the current Bluebird Gap Farm property, and Captain John Sipsey bought the other half. Sipsey willed it to his grandson John Fulcher in 1701. Ralph Barlowe married Katherine West—of the West family from the Westover Plantation section in Chapter 5—in 1651. When Barlowe died a year later, he left the property to Katherine. She remarried twice, and her descendants include the Wests, the Croppers and the Scarboroughs. The property remained in the West family throughout the 1700s. It was described as 150 acres in Elizabeth City, "northwest upon Noswattock River from Arokoko Creek mouth bound west, southwest with said creek, northeast with Mansotanzick Creek."

The West and Woutten families were tied by marriage, and one of the earliest graves at the Davis Cemetery belongs to Hadrach Woutten, who was born in 1797 and died in 1835. Woutten's wife, Margaret, was a descendant of Katherine West, and she went on to remarry after her first husband's death. Margaret had a daughter, Alice, born in 1851, and her ancestry can be traced through the Davis Cemetery tombstones. Alice married Philip P. Davis, and they had a daughter Beula, the eldest of their many children. Beula married a Mr. Curtis, of whom little is known, and they had a daughter named Lilian. Lilian went on to have a daughter of her own, whom she named Beulah. Lilian's mother, Beula, died in 1913, only four days before her father, Philip, who is believed to have died from a broken heart because "his daughter's untimely demise weighed heavily upon him." Philip had fought with the Old Dominion Dragoons during the Civil War as one of their "bravest and best soldiers." After the war, Philip returned to his birthplace to become a farmer and was described in his obituary as being "highly successful," "a man of high character" and "a widely popular man." Philip's wife, Alice, died in 1922 and was survived by eight children, six grandchildren and a great-granddaughter.

This covers the genealogy and history of the property, and all of this information was passed along to the farm and then was published on its website since its history had been previously unknown. But this information doesn't explain how the cemetery came to be on the property of Bluebird Gap Farm in the first place. Nor does it explain the origins of the 1908 stone base inscribed with a Masonic emblem at the cemetery. Likewise, the identity of the lone figure who used to visit the cemetery on a yearly basis remains a mystery.

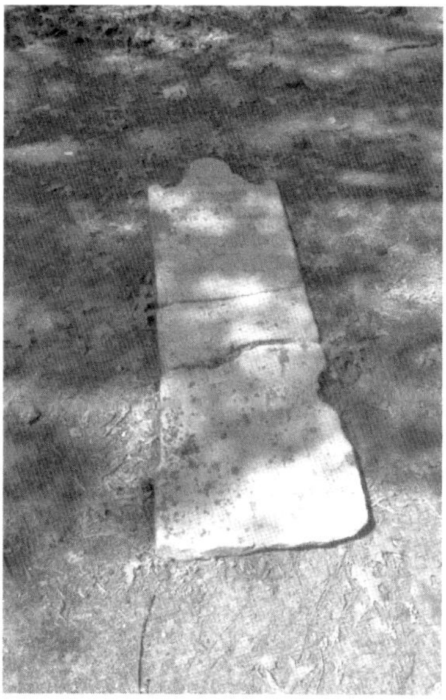

A tombstone at the Davis Cemetery. *Photo by author.*

However, we do know that the property remained in the Davis family throughout the early 1900s, until the Warren family bought it around 1950. They were the last private owners before the City of Hampton acquired it and opened Bluebird Gap Farm in 1969. The farm had previously been opened at a different location under the name of Old MacDonald's Farm but was forced to move when the Coliseum construction started.

City employees and farm workers have always known about the old cemetery, but it wasn't advertised since it was located in a place designed for children and families. It is historically noteworthy, though, and far away from the animal exhibits and playground. It is quite a peaceful spot during the day, but after dusk, it tells a different story. A number of strange incidents have occurred in and around the old Davis Cemetery, from cold spots and creepy chills to unexplained lights flickering after dark. Employees have reported hearing strange sounds and seeing misty figures in the cemetery on various occasions. One former employee claimed that his work glove moved by itself on the truck seat beside him while he was working at the cemetery. Those are only a few of the supernatural sightings, but probably even more would be reported if more people knew about the cemetery and bothered to visit it.

ST. JOHN'S EPISCOPAL CHURCH

We will leave the Davis Family Cemetery and Bluebird Gap Farm to head to our next stop, and next cemetery, in Hampton. The grave at this cemetery is also old, dating to 1700, and is located next door to 892 West Pembroke Avenue, which is right off I-64 and very close to Hampton University. The grave we are visiting belongs to a relatively unknown man by the name of Peter Heyman, located at the site of the "New Church of Kecoughtan," or the Third Church of Elizabeth City Parish, built on Pembroke Plantation. Four different churches had to be built during the 1600s and 1700s as Hampton's city limits were incorporated. The four different parish sites are all part of the present-day St. John's Episcopal Church. The first church was built in 1610, making it the oldest English-speaking church in America, near present-day La Salle and Chesapeake Avenues. The second church was built in 1623 on the grounds of present-day Hampton University.

The third church, and the site we are now visiting, was built in 1667 and was in use until 1728. The original foundations of this church still remain and are marked off by a border of bricks. A low brick wall also surrounds the entire site. Inside the wall lies Peter Heyman's grave; he was an intellectual, land-loving man who was senselessly killed by pirates in 1700.

Piracy was so out of control by the mid-1600s that the British government sent a fleet of ships to guard the Chesapeake Bay against pirates. By 1700, piracy still ran rampant in the southeastern ports of Virginia, despite the presence of the official British ships. Louis Guittar was one of the most notorious pirates of the time, predating Blackbeard by a good thirty years, and the British fleet was no deterrent.

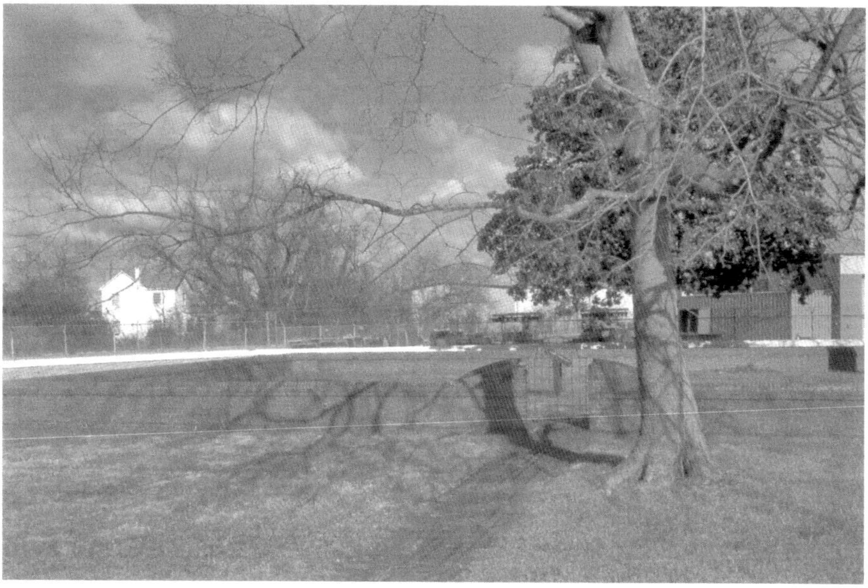

The site of the Third Church of Elizabeth City Parish. *Photo by author.*

Captain Guittar liked to hide along the Eastern Shore of North Carolina to attack ships as they entered the Chesapeake Bay, mostly stealing tobacco, hemp and slaves. James Beard, aka Blackbeard, was from North Carolina and grew up hearing tales of the pirate Guittar; judging by Blackbeard's (mis)conduct in later years, the tales seemed to have made quite an impression on him.

Unlike Blackbeard, little is known about Guittar and his appearance, other than the fact that he was French, had a milky eye, was scarred by smallpox and wore a gold toothpick around his neck. He was from the Caribbean islands and was mostly active in the Caribbean and West Indies during the 1690s. Then he discovered the wealth that American ports, and Virginian in particular, had to offer.

Piracy was such a threat by 1700 that the British admiralty sent the twenty-eight-gun frigate HMS *Shoreham* with a crew of 128 men to southeastern Virginia. The *Shoreham* was greeted at the Chesapeake Bay harbor by the sight of Guittar attacking five ships. The pirate even allowed his crew to hack one of the smallest ships to bits just for fun before chasing the remaining ship, the *Quaker*, into Lynnhaven Bay. The *Shoreham* docked long enough to allow a civilian group of volunteers to board the ship, and then it bore down on Guittar under the direction of Captain William Passenger.

The *Shoreham* was finally able to run the pirate ship ashore, forcing Guittar and his surviving crew members to surrender. Guittar was shipped back to England separately from his crew. He begged for the king's mercy and was promptly hanged instead. Guittar's ship, *Le Paix* (meaning "peace"), was ordered to be sold, but after sitting idle for a year, it was used to transport tobacco from Jamestown to England. A total of 26 pirates died during that daylong battle, and 124 surrendered and were shipped to England as prisoners, with 15 dying en route.

Before Guittar technically surrendered, three members of his crew were captured on land, so they weren't included in the terms of the surrender. These three were ordered by Governor Nicholson to be hanged at three different ports along the Chesapeake Bay as a warning to deter other pirates from invading. The three managed to escape but were soon found in Accomack County. They were returned to Hampton where, under the governor's orders, they were hanged at different ports until "they rott and fell away." It is believed that these pirates are the very same pirates who were eventually found buried under the cellar at Carter's Grove, mentioned in Chapter 3.

Only one fatality was recorded on board the *Shoreham* that momentous day: Peter Heyman. Heyman had been appointed deputy postmaster general in 1692 for all of the plantations in Maryland and Virginia. Deputy Postmaster General Heyman single-handedly set up the first operational postal system in America. In 1698, the English commissioners of the customs made Heyman the collector of customs of the Lower James River District. Heyman seems to have spent most of his time on dry land, in an office or on the docks, and was certainly no sea captain or sailor.

Perhaps Heyman was looking for adventure in his mundane life on the day of the battle because he willingly volunteered to board the *Shoreham* to go off and fight pirates once he heard about Guittar's antics. He fought alongside Lieutenant Governor Frances Nicholson on the quarterdeck of the *Shoreham* until Heyman was shot in the chest by a pirate's lucky bullet. It was Governor Nicholson himself, the very same one who helped fund Grace Episcopal Church in Yorktown, who ordered Heyman's tombstone. The inscription reads:

> *In memory of Peter Heyman, Esq., grandson of Sir Peter Heyman of Summerfield, in the county of Kent (England). He was Collector of Customs in the lower district of the James River and went voluntarily on board the King's ship "Shoreham" in pursuit of a pirate who greatly*

> *infested the coast. After he had behaved himself seven hours with undaunted courage, was killed with a smallshot the 29th day of April, 1700. In the engagement he stood next to the Governor, upon the quarterdeck, and he was here honorably interred by his order.*

Peter Heyman's final resting place lies on a lonely stretch of road with few buildings around. It's certainly not a busy thoroughfare, and once again, you could drive right past it and never know what you were missing. Heyman, who so gallantly proved his worth during life, seems to have been utterly forgotten in death. Even the inscription on his tombstone at the site of the Third Church of Elizabeth City is no longer legible.

The fourth and last church, St. John's Episcopal Church, was built in 1728 and is located in downtown Hampton at 100 West Queen's Way. It was damaged, along with most of the city of Hampton, during the War of 1812 and then again during the Civil War. Confederate soldiers even destroyed the church bell and set fire to the building to prevent it from falling into Union hands. Much of the building was badly burned, but the four

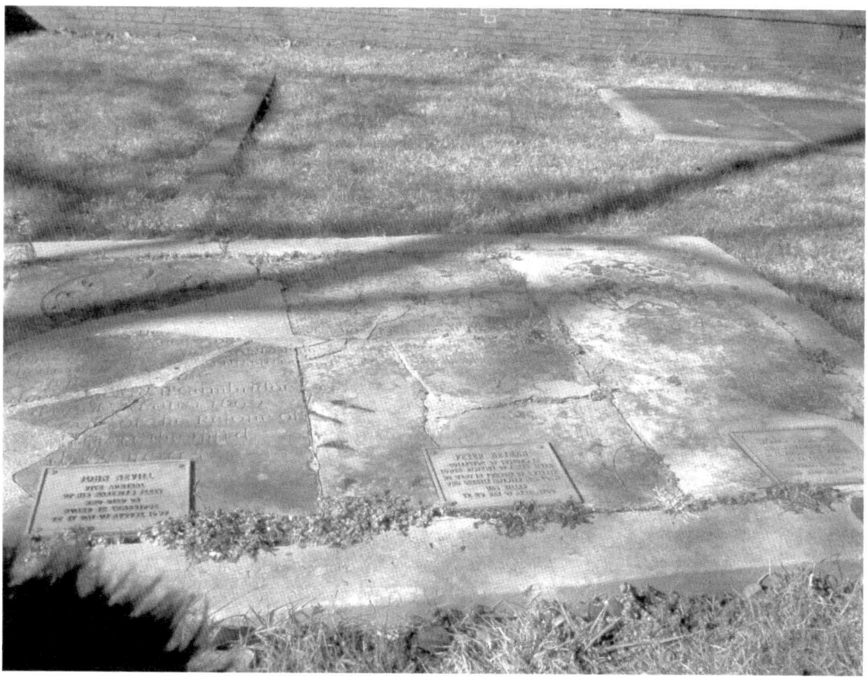

Peter Heyman's grave. *Photo by author.*

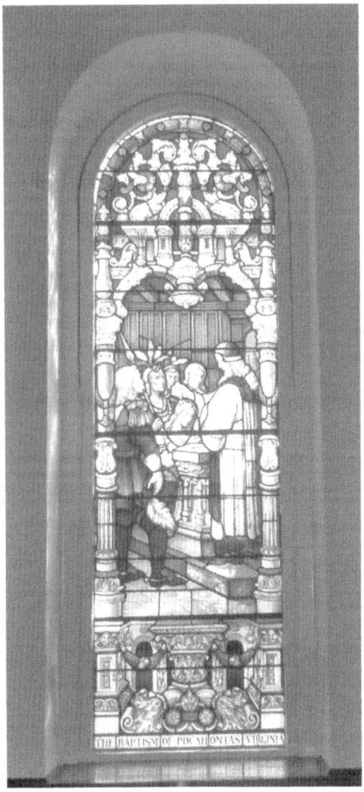

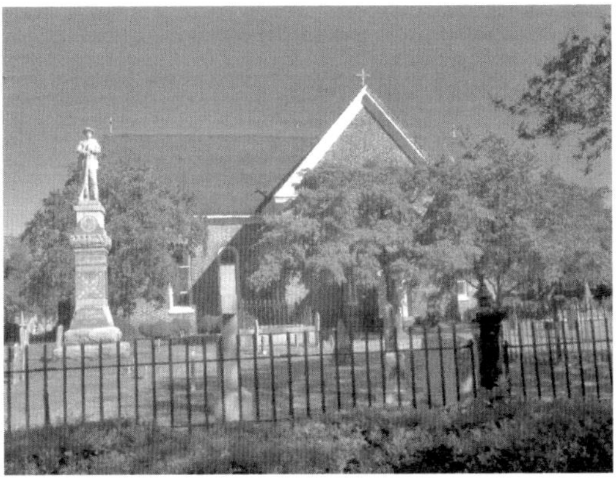

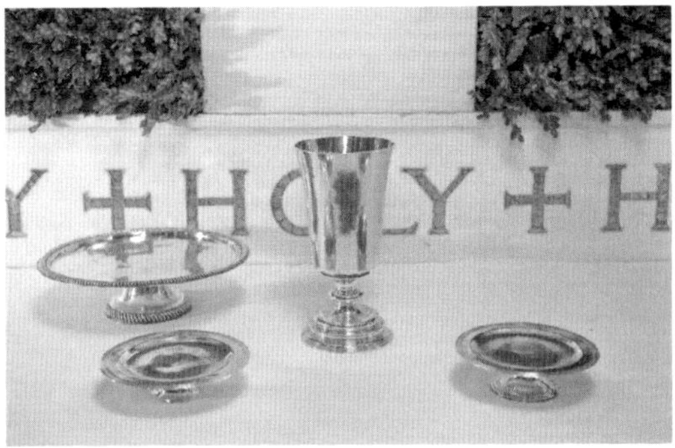

Above, left: A stained-glass window of Pocahontas. *Photo by author.*

Above, right top: St. John's Episcopal Church. *Photo by author.*

Above, right bottom: Holy Communion silver. *Photo courtesy of St. John's Episcopal Church.*

main walls survived, blackened though they were, making it the only existing colonial structure in downtown Hampton today.

Besides the historical value of the church itself, it features a memorial to the first surviving English baby born in America, Virginia Laydon, which is located next to the south wall of the church. Laydon and her parents were members of the first church, and the locations of their exact graves are unknown, so a memorial was erected in their honor. The church also features an 1887 stained-glass window of Pocahontas's baptism. Most noteworthy is the 1618 communion silver housed in a vault inside the church. The first church was destroyed during the Indian Massacre of 1622, but the silver was smuggled out of the church to Jamestown by Governor Yeardley. Currently, the Holy Communion silver is still used for special occasions in the only remaining St. John's Episcopal Church.

FORT MONROE

The next stop on our Hampton Roads tour is also, sadly enough, our last one. Fort Monroe was the first military base established in America, and unfortunately, it closed in 2011. However, the Casemate Museum located inside the fort is still open to the public, along with the pet cemetery atop it. The English originally inhabited the area between 1607 and 1609, calling it Old Pointe Comfort. The first fort, Fort Algernourne, was erected in 1609 by Captain John Ratcliffe, whom you briefly met as he usurped President Wingfield and was tortured by Indians in Chapter 3. The name changed from Fort Algernourne to Fortress Monroe. In 1834, the current star-shaped fort was built and called Fort Monroe.

The fort has changed a lot over the years, with the original one being damaged by a hurricane in 1667 and then rebuilt only to be destroyed by another hurricane in 1749. In 1781, during the Revolutionary War, the Patriot army used the ruins of the fort as its base of operations. After the war, it fell into disuse for the next thirty years.

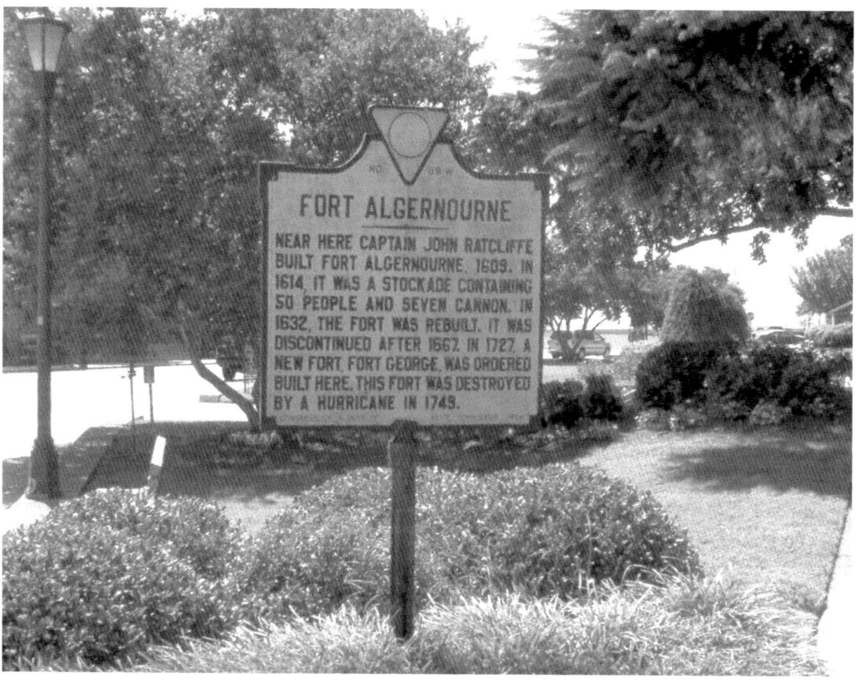

Fort Algernourne's sign. *Photo by author.*

After the War of 1812—which destroyed the entire city of Hampton, including what remained of Fort Monroe—it was decided that the fort needed to be reestablished since it offered a perfect vantage point from which to watch for incoming ships. Perhaps if the fort had been operational prior to the War of 1812, the city of Hampton could have been saved, and the entire war could have been prevented. Construction on the new fort finally began in 1819 and was finished in 1834. The new fort was named after former U.S. president James Monroe. The fort is one of the few in America to be completely surrounded by a moat of seawater. During its final construction, a Masonic lodge was installed inside the fort, but since then it has been removed from government property.

Fort Monroe was fully operational throughout the Civil War, with the Federal government constantly keeping Union soldiers there until 1865. President Abraham Lincoln held meetings at Fort Monroe and in ships off its coats during the war. In fact, President Lincoln held an informal peace meeting with Confederate vice-president Stephens at Fort Monroe shortly before the end of the war. Also, an approximate ten thousand escaped slaves, called "contrabands," sought refuge at Fort Monroe from 1861 until 1865.

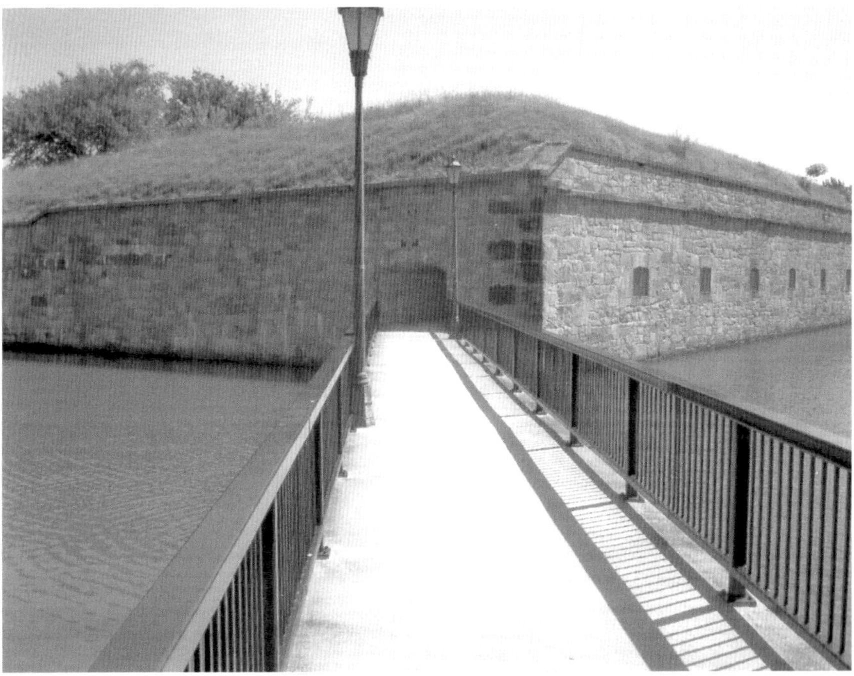

The casemate entrance. *Photo by author.*

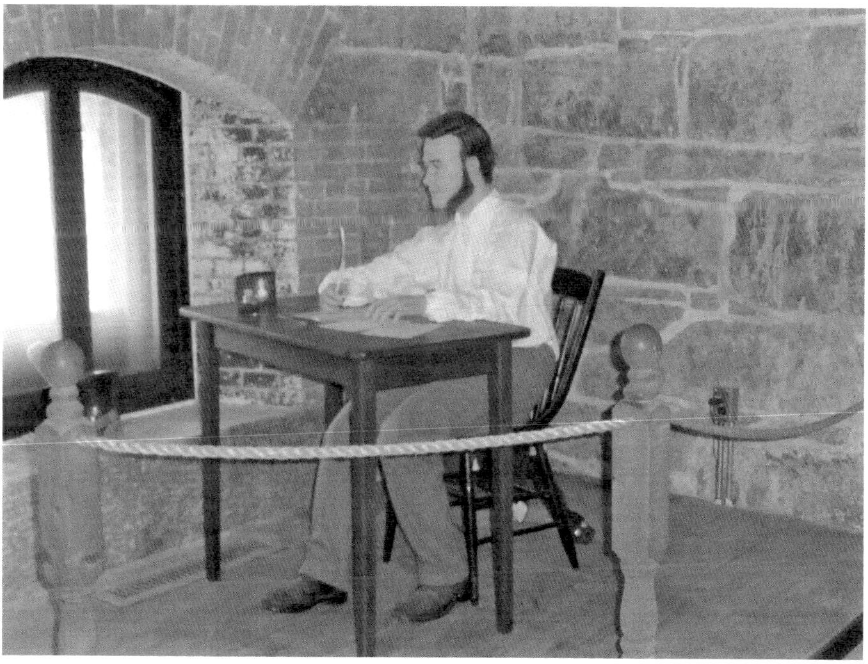

Edgar Allan Poe's room. *Photo by author.*

Fort Monroe's walls have been home to a couple of famous people, like Edgar Allan Poe, who wrote *The Cask of Amontillado* while he was temporarily stationed there. Of course, Confederate president Jefferson Davis was imprisoned there after the Civil War ended. President Davis was forced to flee from his Richmond home on April 2, 1865, when the city was evacuated. Davis and his family fled to Danville and then continued to head west. Davis was captured in Georgia on May 10, 1865, and was charged with treason and conspiracy to commit murder against President Lincoln. Supposedly, Davis was on his way to Texas, where he planned to start a new confederacy and continue the war. Instead, he was escorted back to Virginia and placed in solitary confinement in Casemate #2 at Fort Monroe. Currently, his former cell, along with Poe's former room, is part of a museum open for public tours.

Ex-president Davis was kept shackled in the cell, his health ailing, until his wife, Varina, came to his aid, bringing along a physician to examine him. Davis nearly died in Varina's arms but survived and was released two years after his capture, in May 1867. He traveled to Europe before retiring in Mississippi and died there of natural causes at the age of eighty-one in 1889.

He was eventually buried at the Hollywood Cemetery, 412 South Cherry Street in Richmond, near his son.

Davis and his family lived in the White House of the Confederacy on Clay Street in Richmond from 1861 until 1865. While living there, Davis's five-year-old son, known as "Little Joe," climbed onto a patio railing, lost his footing and fell fifteen feet headfirst onto the brick pavement below. He died within minutes of landing, and days later, Davis had the patio torn down. Officials at the Hollywood Cemetery donated a free burial plot for Little Joe, and the boy's friends collected forty dollars to buy a monument in his honor. Later, a statue of Confederate president Davis was erected near Little Joe's tomb. After Davis's death in 1889, he was first buried in New Orleans, and then Varina had his body moved to the Hollywood Cemetery in 1893. Little Joe's body was moved to be closer to his father's grave.

President Davis's ghost doesn't reside at Fort Monroe, but a number of other ghosts supposedly haunt Fort Monroe today. There is everything from Confederate soldiers to army personnel of yesterday. It is not such a stretch to believe in the ghosts once you see the casemate and "Ghost Alley" for

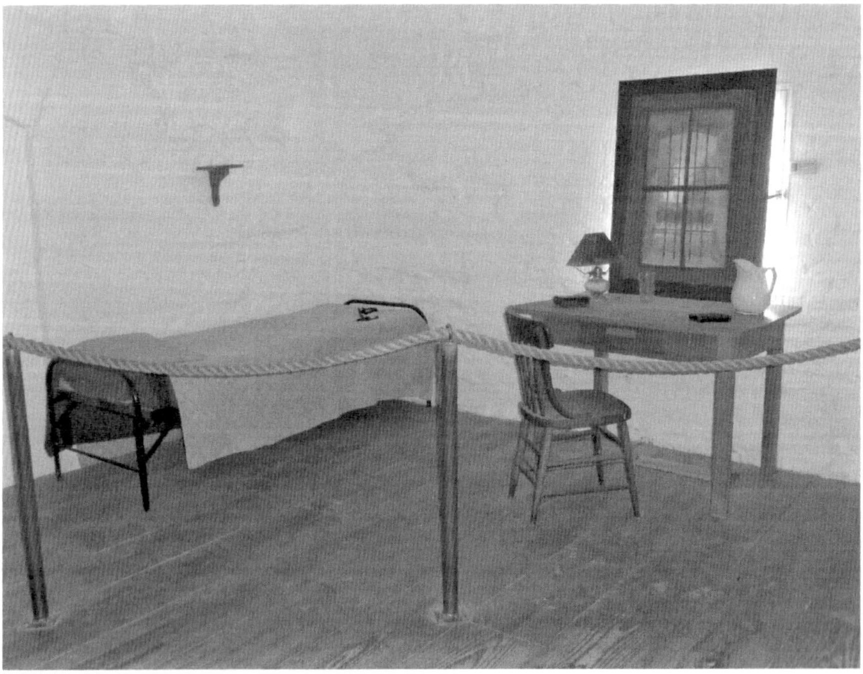

Confederate president Davis's cell. *Photo by author.*

yourself. One of the most famous ghosts at Fort Monroe is the "Woman in White" who roams an area known as Ghost Alley.

The tale of the Woman in White is a tragic love story of the twentieth century. A young couple was stationed at Fort Monroe when the husband was deployed, leaving his wife behind in Virginia. But she didn't remain alone for long, quickly taking a lover. The woman's husband caught them in the act when he received a weekend furlough and came home unexpectedly. The husband proceeded to chase the lover away and then turned around and shot his cheating wife dead. Her spirit lingers, though, scantily clad in the lacy white negligee she died in, and she has been dubbed the Woman in White because of her attire.

AFTERWORD

Now you have met more than one ghost and heard a host of other supernatural tales that were as haunting as the facts were historically accurate. Hopefully, there has been something along our tour to appeal to everyone: historians and curiosity-seekers, ghost and witch hunters, conspiracy theorists and skeptics and psychics and paranormal investigators alike. You have gotten to read some poetic epitaphs at unexplored cemeteries, search for buried pirate treasure and cross the thresholds of some of the oldest houses and churches in America.

You have seen for yourself that the Hampton Roads area of Virginia truly is America's birthplace, with a number of firsts happening here. You have visited the first American business, permanent settlement and military fort. You have seen the first mental institution, Anglican church and Potter's Field. You have learned about the first Native American reservations, American fairy tale, postal system and conspiracy theory. You have met the first baby born in America and the first German immigrants, and you have gotten to tour the first and oldest African American cemetery in America.

But alas, we have come full circle and to the end of our tour of Hampton Roads. You can look across the Chesapeake Bay and see the Southside or north to Blackbeard's Point or inland to downtown Hampton where the Blackbeard Festival is held every summer. The festival really shouldn't be missed, and there are countless other places of interest that we skipped, like Phoebus, Causey's Mill in Newport News and many, many more…but they will have to wait until our next Hampton Roads tour of curiosities.

BIBLIOGRAPHY

Behrend, Jackie Eileen. *Ghosts of America's East Coast*. Birmingham, AL: Crane Hill Publishers, 2001.

———. *The Hauntings of Williamsburg, Yorktown, and Jamestown*. Winston-Salem, NC: John F. Blair, 2000.

Boyer, Paul S., Clifford E. Clark, Joseph F. Kett, Neal Salisbury, Harvard Sitkoff and Nancy Woloch. *The Enduring Vision*. Vol. 1. Boston: D.C. Heath and Company, 1996.

Campbell, George S., A. Lawrence Kocher and Howard Dearstyne. "Architectural Report: Public Magazine." Colonial Williamsburg Foundation. http://research.history.org/DigitalLibrary/view/index.cfm?doc=ResearchReports\RR0145.xml&highlight=0145.

Chewning, Alpheus J. *Haunted Virginia Beach*. Charleston, SC: The History Press, 2006.

Cobb, J. Michael, Edward B. Hicks and Wythe Holt. *Battle of Big Bethel: Crucial Clash in Early Civil War Virginia*. El Dorado Hills, CA: Savas Beatie LLC, 2013.

Dabney, Virginius. *Virginia: The New Dominion*. New York: Doubleday & Co., 1971.

Duell, Prentice, and Herbert S. Ragland. "Governor's Palace Archaeology Report, Block 20 Building 3A." Colonial Williamsburg Foundation. http://research.history.org/DigitalLibrary/view/index.cfm?doc=ResearchReports\RR0093.xml&highlight=0093.

Editors of *American Heritage Magazine*. *A Guide to America's Greatest Historic Places*. Rockville, MD: American Heritage, 1985.

Eighmey, Kathleen M. *The Beach, A History of Virginia Beach, Virginia*. Virginia Beach, VA: Department of Public Libraries, 1976.

Erickson, Mark St. John. "Hampton Hangs Two Members of Blackbeard's Crew 295 Years Ago." *Newport News (VA) Daily Press*, January 27, 2014.

Fredericksburg (VA) Freelance Star. "Church Doesn't Dig Cemetery Excavation." September 17, 1991.

Godson, Susan H., ed. *A Guide to the Memorials of Bruton Parish Church*. Williamsburg, VA: Bruton Parish Church, 2006.

Goodwin, Bill. *Frommer's Virginia*. Hoboken, NJ: Wiley Publishing Co., 2006.

"Grace Sherwood: The Virginia Witch." *William & Mary College Quarterly* (1895).

Grassl, Gary Karl. *First Germans in America*. Washington, D.C.: German Heritage Society of Greater Washington, D.C., 1997.

Hakluyt, Richard. "Instructions by Way of Advice, for the Intended Voyage to Virginia." National Humanities Center. http://nationalhumanitiescenter.org/pds/amerbegin/settlement/text4/text4read.htm.

Hall, Marie Bauer. *Foundations Unearthed*. Lancaster, PA: Veritas Press, 1974.

Hauck, Dennis William. *Haunted Places: A National Directory*. New York: Penguin Group, 2002.

Hodapp, Chris. "The Bruton Vault Story." Lexington, MA: National Heritage Museum Archives, October 2008.

Holler, Anne. *Pocahontas Powhatan Peacemaker*. New York: Chelsea House Publishers, 1993.

Holtzclaw, Mile. "Richly Famous Tombs." *Newport News (VA) Daily Press*, October 22, 2006.

Hudson, J. Paul. *A Pictorial Story of Jamestown, VA*. Richmond, VA: Garrett and Massie, Inc., 1957.

Hume, Ivor Noel. "Excavations at the President's House College of William and Mary Archaeological Report, Block 16 Building 2." Colonial Williamsburg Foundation. http://research.history.org/DigitalLibrary/view/index.cfm?doc=ResearchReports\RR0075.xml&highlight=0075.

Humelsine, Carlisle H. *Williamsburg Tells It Like It Was: The President's Report, 1967*. Williamsburg, VA: Colonial Williamsburg, 1968.

Kelso, William M. *Jamestown the Buried Truth*. Charlottesville: University of Virginia Press, 2006.

Kirkham, Jean, and Debra Boyce. *In Every Generation*. Farmingdale, NY: Minuteman Press, 1997.

Klingelhofer, Eric. "Peyton Randolph House Archaeological Report, Block 28 Building 6 Lot 236." Colonial Williamsburg Foundation. http://research.

history.org/DigitalLibrary/view/index.cfm?doc=ResearchReports\RR1540.xml&highlight=1540.

Kocher, A. Lawrence. "Benjamin Waller House Architectural Report, Block 1 Building 16." Colonial Williamsburg Foundation. http://research.history.org/DigitalLibrary/view/index.cfm?doc=ResearchReports\RR1002.xml&highlight=1002.

Lee, Lauranett L. *Making the American Dream Work: A Cultural History of African Americans in Hopewell, Virginia*. Hampton, VA: SPS Press Publishing, 2008.

McCary, Ben. *Indians of 17th Century Virginia*. Williamsburg: Virginia 350th Anniversary Celebration Corp., 1957.

McKenzie, Bryan. "Owner of Carter's Grove Navigates Financial Troubles." *Newport News (VA) Daily Press*, January 6, 2010.

McMichael, George, ed. *Anthology of American Literature*. Vol. 1. 6th ed. Upper Saddle River, NJ: Prentice-Hall, Inc., 1997.

Miles, Barry W. *Cemeteries of the City of Hampton*. Berwyn Heights, MD: Heritage Books, 2007.

Miles, Barry W., and Gertrude Stead. *Cemeteries of the City of Newport News, Formerly Warwick County, Virginia*. Berwyn Heights, MD: Heritage Books, 2009.

New York Times. "Famous Old Nelson House." February 18, 1894.

Norman, Michael, and Beth Scott. *Historic Haunted America*. New York: Tor, 1995.

"Old Kecoughtan, Elizabeth City County, Virginia." *William & Mary College Quarterly* 9, no. 2 (n.d.).

O'Neill, Helen. "History and Faith: Tracing the First Black Family in America." Associated Press, February 1, 1998.

Parker, Stacy. "Community Celebrates Opening of Renovated Cayce Hospital." *Virginian-Pilot*, July 7, 2014.

Ragland, Herbert S. "James Geddy House Archaeological Report, Block 19 Building 11." Colonial Williamsburg Foundation. http://research.history.org/DigitalLibrary/view/index.cfm?doc=ResearchReports\RR1447.xml&highlight=1447.

———. "King's Arms Tavern Archaeological Report, Block 9 Building 29A & B Lot 24." Colonial Williamsburg Foundation. http://research.history.org/DigitalLibrary/view/index.cfm?doc=ResearchReports\RR1158.xml&highlight=1158.

———. "Robert Carter House Historical Report, Block 30–2 Building 13 Lot 333–336." Colonial Williamsburg Foundation. http://research.history.org/DigitalLibrary/view/index.cfm?doc=ResearchReports\RR1610.xml&highlight=1610.

"Roadside Memorial Pose Hazard." *Newport News (VA) Daily Press*, May 29, 1994.

Rountree, Helen C. *Pocahontas' People*. Norman: University of Oklahoma Press, 1990.

Rountree, Helen C., and E. Randolph Turner III. *Before and After Jamestown*. Gainesville: University Press of Florida, 2002.

Rountree, Helen C., and Thomas E. Davidson. *Eastern Shore Indians of Virginia*. Charlottesville: University of Virginia Press, 1997.

Rouse, Parke, Jr. *The Good Old Days in Hampton and Newport News*. Petersburg, VA: Dietz Press, 1986.

———. *Jamestown's Story*. Petersburg, VA: Dietz Press, 2006.

———. *Virginia, A Pictorial History*. New York: Charles Scribner's Sons, 1975.

Ruegsegger, Bob. "Local Ghosts and Witches Rise from New Book." *Virginian-Pilot*, October 24, 2004.

Schlesinger, Catherine Savedge. "Wren Building at the College of William and Mary Restored: Summary Architectural Report of Interior Restoration 1967–1968 Block 16 Building 3." Colonial Williamsburg Foundation. http://research.history.org/DigitalLibrary/view/index.cfm?doc=ResearchReports\RR0194.xml&highlight=Wren%20Chapel.

Smith, Terence. "Beyond Jamestown." *Smithsonian Magazine* (May 2007).

Stephenson, Mary A. "Ludwell Mansion (NB) Historical Report, Block 28 Building 4A Lot 234." Colonial Williamsburg Foundation. http://research.history.org/DigitalLibrary/view/index.cfm?doc=ResearchReports\RR1528.xml&highlight=1528.

Taylor, L.B., Jr. *Civil War Ghosts of Virginia*. Lynchburg, VA: Progress Printing Co., Inc., 1995.

———. *The Ghosts of Virginia*. Vol. 3. Lynchburg, VA: Progress Printing Co., Inc., 1996.

———. *The Ghosts of Williamsburg*. Lynchburg, VA: Progress Printing Co., Inc., 2001.

———. *The Ghosts of Williamsburg and Nearby Environs*. Lynchburg, VA: Progress Printing Co., Inc., 1983.

Tennant, Diane. "Researchers Drill Lynnhaven House for Its History." *Virginian-Pilot*, December 7, 2007.

Tucker, George Holbert. *Virginia Supernatural Tales*. Virginia Beach, VA: Donning Company, 1977.

Turnage, Sheila. *Haunted Inns of the Southeast*. Winston-Salem, NC: John F. Blair, 2001.

Virginian-Pilot. "Use Sense, Not Science, for Shore Drive Speeds." January 2, 2006.

Von Toal, Dagmar, and Michael Jarvis. "Dagmar von Toal's 1992 Excavation Report." Colonial Williamsburg Foundation http://research.history.org/DigitalLibrary/View/index.cfm?doc=ResearchReports\RR1716.xml.

Washington, Jim. "Interactive Map: Drowning Deaths at Norfolk, Virginia Beach Beaches Since 1998." *Virginian-Pilot*, September 26, 2008.

Woolley, Benjamin. *Savage Kingdom*. New York: HarperCollins, 2007.

Yetter, George Humphrey. "When Blackbeard Scourged the Seas." *Colonial Williamsburg* 15, no. 1 (Winter 1992): 22–28.

———. *Williamsburg Before and After*. Williamsburg, VA: Colonial Williamsburg Foundation, 1988.

Zacks, Richard. *The Pirate Hunter*. New York: Hyperion, 2002.

INDEX

Y

ABOUT THE AUTHOR

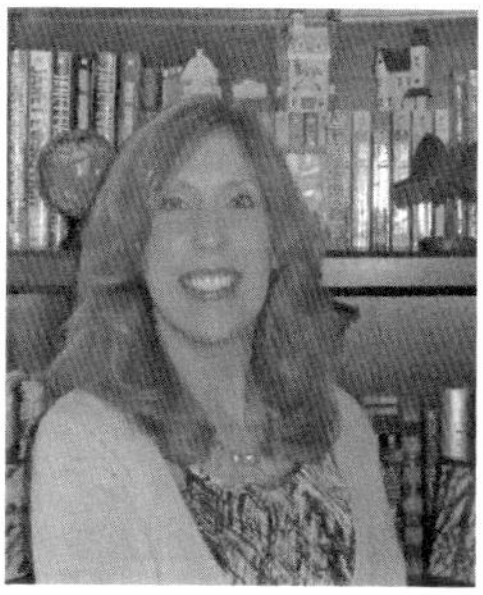

Tamy Kay Thompson attended the Art Institute of Pittsburgh and took correspondence courses through the Institute of Children's Literature. She has created pop-up books, short stories and nonfiction articles that have been published in *Fun for Kidz* magazine. Thompson is a southeastern Virginia historian who moved to the area in 1991. Since then, she has been researching and collecting information for *Curiosities of Hampton Roads*, her first nonfiction book for adults. Currently, she lives in Newport News with her husband and two children.